MULTI-STAGE SPACE GUNS, MICRO-PULSE NUCLEAR ROCKETS, AND FASTER-THAN-LIGHT QUARK-GLUON ION DRIVE STARSHIPS

STEPHEN BLAHA, PH. D.

BLAHA RESEARCH

ISBN: 978-0-9845530-7-5

Cover Credits
A symbolic depiction of various starships powered by quark-gluon thrust generated from the collision of particle beams. Cover by Stephen Blaha © 2012.

Rev. 00/00/01   February 25, 2013

To My Wife Margaret

Once more unto the breach, dear friends!
King Henry V
Will Shakespeare

Some Other Books by Stephen Blaha

From Asynchronous Logic to The Standard Model to Superflight to the Stars (Blaha Research, Auburn, NH, 2011)

From Asynchronous Logic to The Standard Model to Superflight to the Stars volume 2: Superluminal CP and CPT, U(4) Complex General Relativity and The Standard Model, Complex Vierbein General Relativity, Kinetic Theory, Thermodynamics (Blaha Research, Auburn, NH, 2012)

The Algebra of Thought & Reality: The Mathematical Basis for Plato's Theory of Ideas, and Reality Extended to Include A Priori Observers and Space-Time; Second Edition (Pingree-Hill Publishing, Auburn, NH, 2009)

The Origin of the Standard Model: The Genesis of Four Quark and Lepton Species, Parity Violation, the ElectroWeak Sector, Color SU(3), Three Visible Generations of Fermions, and One Generation of Dark Matter with Dark Energy (Pingree-Hill Publishing, Auburn, NH, 2007)

Physics Beyond the Light Barrier: The Source of Parity Violation, Tachyons, and A Derivation of Standard Model Features (Pingree-Hill Publishing, Auburn, NH, 2007)

Quantum Theory of the Third Kind: A New Type of Divergence-free Quantum Field Theory Supporting a Unified Standard Model of Elementary Particles and Quantum Gravity based on a New Method in the Calculus of Variations (Pingree-Hill Publishing, Auburn, NH, 2005)

Quantum Big Bang Cosmology: Complex Space-time General Relativity, Quantum Coordinates, Dodecahedral Universe, Inflation, and New Spin 0, ½, 1 & 2 Tachyons & Imagyons™ (Pingree-Hill Publishing, Auburn, NH, 2004)

SuperCivilizations: Civilizations as Superorganisms (McMann-Fisher Publishing, Auburn, NH, 2010)

Standard Model Symmetries, And Four And Sixteen Dimension Complex Relativity; The Origin Of Higgs Mass Terms (Blaha Reasearch, Auburn, NH, 2012)

Available on bn.com, Amazon.com, Amazon.co.uk and other international web sites as well as at better bookstores (through Ingram Distributors).

Preface

NASA, and its predecessor the National Advisory Committee for Aeronautics (NACA), and European and Asian Space Programs have made magnificent efforts to begin the exploration of space. In recent years this effort has been severely budget constrained due to enormously increasing governmental deficits and a decline in the standard of living in the Western world in particular. NASA is beginning massive layoffs of skilled personnel that will total approximately 1,000,000 workers by 2014. These budget declines and constraints make it imperative that the component projects of the various space programs be prioritized for maximum results with restricted budget monies.

This book proposes a long-term space program to put Man in space in an aggressive, cost-effective way.

Currently there are two general types of programs underway in the various space programs: near earth space projects such as the placement of communication satellites in orbit around the earth, and scientific exploration efforts further from earth reaching to the outer solar system. Near earth projects such as the placement of communication satellites are economically viable. It would be economically beneficial to develop more cost effect methods for launching near earth satellites and other cargo. While several alternative private space launch programs are underway, they all rely on rocketry. This book proposes an alternative approach that appears to be much less expensive using a new concept: *multi-stage space guns* to shoot cargo into space up to 150 kilometers or so. Our concept of a multi-stage space gun enables larger payloads to be sent into near space more efficiently (with less propellant consumption) than a conventional gun such as the "Big Bertha" gun of World War I that sent shells up to 80 – 90 miles into Near Space. It takes one kilogram of propellant to send one kilogram of cargo up 150 kilometers. In contrast it takes a massive rocket much fuel to perform the same chore. The economics is clear. Rockets could still be used to transport delicate cargo into orbit. But a multi-stage space gun could do the workhorse transport.

Scientific exploration efforts in the wider solar system have also major engineering and economic challenges. A major thrust of scientific efforts has been the search for life on other planets and moons. Another aspect of these efforts is the determination of the physical characteristics of various planets and moons. For example, do various Jovian and Saturnian moons have a liquid ocean within and, if so, what are its characteristics?

If space programs had generous, or unlimited, budgets then these scientific efforts would be sensible and meritorious. However, in the present world economic climate it is clear that space programs must be carefully prioritized to forge the most economical, *long-range* space exploration program.

This author believes that NASA has not pursued the most economical long range program because of its commitment to chemical rockets. The commitment to chemical rockets started after World War II. Building larger and larger chemical rockets is a major factor in the expensive NASA program that has evolved over the past sixty years. Currently there is a proposal before NASA for a giant ½ billion dollar chemical rocket to explore the Solar System.

The Russian space program has followed a similar course with the same negatives. However the Russians have started a nuclear rocket program as an alternative to large chemical rockets.

In the 1960's both the US and Soviet programs explored the possibility of nuclear rockets. They both realized the advantages of nuclear rocket based exploration of the solar system. However, both countries decided that nuclear rockets had the potential for catastrophic results (*and also, paradoxically, might be too successful by leading to a major space race*), and terminated their nuclear rocket programs. Both programs then focused on chemical rockets as safer, known space vehicles despite their extremely high costs.

From a long range, safety-oriented, perspective chemical space vehicles should be used near earth and nuclear rockets should be used for long distance travel within the solar system. Therefore the development of nuclear rockets for long distance space travel should have a higher priority, and higher funding, then chemical rocket exploration of the Solar System. The Solar System will still be about the same in the ten years or so it would take to build a nuclear rocket ship in space. With a nuclear ship the costs of Solar System exploration would be substantially less and the speed of its exploration substantially increased.

In this book we consider two types of nuclear rockets. One type of rocket is designed for use in the solar system. It would have a reactor with a lifetime of 15-20 years before needing replacement. This *"short range" nuclear rocket* would be capable of manned flights throughout the Solar System. Nuclear ships would be far more cost effective and substantially reduce travel times. For example, a nuclear rocket trip to Mars would take six months while a chemical rocket trip would take about two years.

The second type of nuclear rocket would be for use for navigation in the vicinity of other stars' solar systems. Because time "goes quickly" on faster-than-light starships, a starship needs a *"long shelf life" nuclear reactor* system with a "shelf life" lasting from thousands of years to millions of years. Faster than light starship time passes much more quickly than earth time. For example, starship time increases by a factor of 1,000 times earth time on a starship traveling at 1,000 times the speed of light. One year of earth time would amount to 1,000 years of time on the starship. The longer time intervals on fast starships will require crews to travel in suspended animation until they reach their destination.

Long shelf life nuclear reactors require a new design(s) that has hitherto not been developed by nuclear physicists because there was no previous perceived need for such a reactor. Designs for long shelf life nuclear reactors are described later in this book.

Most importantly, the book develops detailed conceptual *starship models* (qualitatively in the text and mathematically in appendices). The starship designs are all based on a new version of The Standard Model of Elementary Particles developed by this author in recent years. An intrinsic part of this model, and the source of many of the unusual features of The Standard Model, is the identification of neutrinos and quarks as faster-than-light (tachyon) elementary particles. This identification "explains" parity violation and the form of the fermion spectrum and much of The Standard Model's form. Hitherto the Standard Model was viewed as experimentally correct but theoretically bizarre. Our derivation (described elsewhere) connects Standard Model features directly to features of complex space-time. Strong evidence already exists for the faster than light neutrinos and is presented in chapter 6.

The key to faster-than-light starships is a faster-than light ion thrust generated by quark-gluon plasmas. Quark-gluon plasmas can travel faster-than-light and provide the thrust needed for faster-than-light starships. There are several possible design variations for faster-than-light starships. One design was presented in previous books by this author. This book presents new designs that appear to be also worthy of consideration.

We propose to spread the starship R&D costs over several decades. Initially earth-based prototypes could be constructed. One suggestion is to build a prototype engine using the LHC particle accelerator. The cost of the initial decade will be fairly low – perhaps four billion dollars per year. The costs only rise substantially when an actual starship is built – presumably in orbit round the earth or the moon. One research possibility is to build major parts of the starship from ceramic materials made from lunar material. Much of the moon surface has relatively high concentrations of tungsten and iron. Ceramics made from lunar material could be used in starship construction. The tradeoff is the cost of transporting materials from earth vs. the cost of building a lunar infrastructure to make starship components.

A NASA Program prioritized for the long term – constructing nuclear rockets to cost-effectively explore the solar system, using multi-stage space guns to lob material into earth orbit at low price, and constructing a faster than light starship – is a successful long-term space program.

The overall cost of this program would appear to be financially competitive with the recently proposed half-billion dollar chemical rockets and their attendant operational costs over the years. Spread over 30 years our proposal is financially practical and positions NASA for an aggressive foray into the future.

Chapters 1 and 12 describe time frames and expected costs of the various phases of our proposal. Despite the need for budget cutbacks in the U.S. this proposal is financially cost-effective, yet aggressive, and can reasonably fit in our cost-constrained financial situation.

CONTENTS

i

iv

FIGURES and TABLES

1. A Rational, Cost-Effective Approach to Space Travel

NASA, and the space programs of other countries, have done an admirable job in Near Space projects such as communications satellites, the space station, and space-based science projects. They have also sent exploration ships to the moon, Venus, Mars, and the outer planets and moons which have generated vast amounts of data that will help us understand planetary dynamics and the specific features of planets and moons. The study of Jovian and Saturnian moons with their possible subsurface oceans and concomitant possibility of alien life has aroused great public and scientific interest.

All of these efforts have been based on the use of chemical rockets. Chemical rockets have the advantage of a well-developed technology that has incrementally advanced since the 1940's. Unfortunately they also have significant disadvantages for long trips to other Solar System planets. The disadvantages are costliness and lengthy travel times. Lengthy travel times create major human factors issues for manned travel to even the closest planets. Issues include radiation, psychological effects of long isolation and the provision of food, water and oxygen for the astronauts.

Recently Russia, and apparently the United States, have again[1] begun to consider the use of nuclear rockets for planetary voyages. Another recent development is the onset of a private rocket launch industry in the United States. The companies in this fledgling industry are committed to the use of various forms of rockets to launch payloads into Near Space – approximately 100 to 150 miles into space.

---

[1] A nuclear rocket program was started and restarted and then abandoned by NASA's predecessor, the National Advisory Committee for Aeronautics (NACA). For example, see "Steady Nuclear Combustion in Rockets" by E. Sänger, Astronautica Acta, I, Fasc. 2 (1955),

It is apparent to this author, and others, that these new efforts are attempts to develop a comprehensive, cost-effective space program for humanity. However, it is the opinion of this author that there are significant possibilities that have been overlooked in the present and emerging space program. The world-wide funding crunch has made it imperative to examine all reasonable, possible ways to develop a world space program in the most cost-effective manner while not sacrificing opportunities for the exploration of the Solar System and beyond. This book proposes a program that will achieve that goal if certain technological challenges are successfully overcome.

Since there are several aspects to space travel from the earth we have chosen to focus on the three major aspects of space travel.[2] We have also chosen not to discuss existing space vehicles, but rather to consider new, potentially types of space transport bearing in mind the current, and probable near term economic problems which make it imperative to maximize the return on investment in the development and use of space transport vehicles.

1.1 General Requirements for Cost-Effective Spaceship Propulsion

There are self-evident requirements for spaceship propulsion methods at the present time and for the foreseeable future:

1. For a given type of route the propulsion method must be timely and not be slow in terms of human life spans.
2. The propulsion method chosen for a particular class of routes should be the most cost-effective method.
3. The propulsion methods currently possible for any set of routes should be evaluated to find the best propulsion method.
4. As our physical and engineering knowledge evolves over the years, reassessments of propulsion methods should be made for each set of routes. The continuing development of new

---

[2] Travel and transport near earth, in the Solar System, and to the stars.

materials and new energy sources makes periodic[3] reassessment of propulsion methods necessary.

Space travel routes can be divided into three types. Each type has one or more preferred methods of travel. We will consider these types of travel routes and their propulsion mechanisms below. Subsequent chapters will make fairly detailed proposals.

1.2 Three Phases of Space Travel

1.2.1 Near Space Propulsion

The first type of route is travel from the surface of the earth to near regions of space, which we can take to be 150 kilometers. Most space activity in the past forty years has concentrated on this phase with communication satellites being a primary purpose. Chemically powered rockets have been the only propulsion method used until the present.

While chemical rockets and shuttles have been very successful in accomplishing their missions, it appears that other propulsion methods may be more cost-effective for Near Space missions. Building and fueling large rockets are very costly.

We suggest a suitably designed *multi-stage space gun* may provide a more cost-effective delivery vehicle for the delivery of most non-fragile cargo to Near Space. We describe this new type of space gun in chapter 3. Its ancestors are the German big guns of World War I. These guns of World War I (the approximately 100 foot long Big Bertha and a larger gun) that bombarded Paris from a distance of 80+ miles sent their 100+ lb shells as high as 80 – 90 miles above the earth to NEAR Space. Their shells had a speed of 1 mile per second as they emerged from the gun's barrel. These single-stage guns had several defects from the point of view of lofting cargo into Near Space.

The "single-stage" German big guns had a number of deficiencies as a method to loft cargo into space. Gun barrels eroded fairly quickly. More importantly their cargo capability did not scale

---

[3] Reassessments should perhaps be made every twenty years.

upwards satisfactorily from one kilogram shells to very heavy shells. A one kilogram payload required a 1 kilogram charge to propel into Near Space. However a much larger payload, for example a 100 kg payload, would require a much larger charge because the charge would not only propel the payload but it would also propel the "unburned" charge above it in the gun thus increasing the required propellant charge significantly.

A single or multi-stage rocket, with or without boosters, uses over 99% of its weight as fuel to send a payload into space. The fuel is, for the most part, used to propel itself (the fuel) off the ground at an ever-increasing speed into space. In contrast, the propellant for a gun propels the projectile and a fraction of the propellant between the point of burn of the powder charge and the projectile. Thus a sufficiently large single-stage gun could efficiently put a payload up eighty miles into Near Space because the propellant propels the payload and the charge, and not fuel or a rocket casing.

However single-stage gun barrel erosion and the propellant required for large cargoes, among other factors, led NACA not to pursue this technology in the 1960's.

In chapter 3 we describe the design of a multi-stage space gun that would avoid the escalation of propellant charge needed for large payloads and reduce the barrel erosion problem. We view multi-stage space guns as a cost-effective solution to send bulk cargo to space while continuing to use rockets for people transport and the transport of delicate equipment. Thus multi-stage space guns could enable the transport of large quantities of materials to space to build a large space station and to build (nuclear) rockets in orbit to avoid potential nuclear catastrophes due to the launch of nuclear rockets from earth. They also can be important in building the infrastructure of a spaceport on the moon and ultimately Mars. The most expensive part of these projects will be the initial stage of transport of massive quantities of materials to Near Space in a cost-effective manner.

1.2.2 Solar System Travel – Nuclear Rockets

A nuclear rocket program was started and then abandoned by one of NASA's predecessors, the National Advisory Committee for

Aeronautics (NACA). It was felt that nuclear rockets were dangerous should they explode on takeoff from earth or from near earth orbit. In view of the Chernobyl nuclear reactor disaster which impacted on much of Europe this decision was prudent.

However if we can develop mass transport of nuclear rocket construction materials to far earth orbit, or better yet to orbit around the moon, such as multi-level space guns would make possible, then the dangers of nuclear rocketry could be reduced to a minimal level. After construction nuclear rocket(s) could be used to explore the Solar System much more quickly and economically than chemical rockets. For example a chemical rocket trip to Mars is estimated to take approximately two years while a nuclear rocket trip would take approximately six months (and perhaps much less if the nuclear rocket was especially powerful.)

The cost of developing a nuclear rocket would be significant. But its use in interplanetary travel would be far more cost-effective then a chemical rocket. An additional economic benefit would be the reusability of the nuclear rocket for repeated voyages. Rather like nuclear submarines they offer great cruising ranges and long range trips. Finally a well-designed nuclear ion rocket could use hydrogen, methane and other gases mined on the asteroids and moons on outer solar system planets as a source of ion particles. Thus a nuclear rocket need not carry large quantities of the propellant as opposed to chemical rockets which generally need to carry large amounts of fuel. Even if chemical fuels are found on distant asteroids they would, most likely, need to bring a large supply of oxygen.

We can only conclude that a top priority nuclear rocket R&D program is needed. Given tight budgets, expenditures on chemical rocket ventures should be diverted to the nuclear rocket program. Quite simply, we want the horse before the buggy. There is no good reason to expend large sums immediately for most of the science driven rocket trips currently planned. The Solar System will not change much in the next ten years. We should divert funds to a workhorse nuclear R&D program and wait until a nuclear rocket(s) can carry out the delayed projects expeditiously and economically. The Russian space program has recognized the importance of nuclear rockets and is actively engaged in their design and development. The US Space Program should pursue a similar course – perhaps in partnership with Russia.

1.2.3 Extra-Solar Travel

There are two major propulsion issues for travel to other stars and galaxies. First a sensible propulsion method must be found that enables faster-than-light travel to stars and galaxies. When a starship reaches a distant star then a secondary propulsion system must be present for travel within the star's Solar System. The secondary propulsion system would be a combination of nuclear and chemical propulsion systems. The nuclear propulsion unit would handle large-scale movement within the remote solar system. The chemical propulsion unit would handle motion in the vicinity of a specific planet or moon. The starship should also have two nuclear shuttles for round trip visits to the surface of earth-like planets and the surface of moons.

1.2.3.1 Sub-Light Starships

Many proposals have been made for sub-light travel to the stars. One type of proposal suggests a nuclear powered ion drive starship. More exotic proposals exist such as solar sails that use starlight to accelerate a starship. All of these sub-light starship proposals require extremely lengthy travel times extending up to many generations. As a result they have a very limited range of nearby stars and secondarily they are not suitable for the important goals of mining, trade and colonization should we find planets/moons with important minerals, or aliens with whom to trade, or planets similar to earth suitable for human colonization.

Thus these types of starships will only be useful for scientific exploration. This is a worthy goal. But the possibilities of the previous paragraph would appear to be far more significant for the long-term benefit of Mankind. We therefore believe sub-light starships are not desirable.

1.2.3.2 Faster-Than-Light Gravity-Propelled Starships

There are a number of proposals for starship propulsion based on gravitation effects. A prominent propulsion proposal is called *Alcubierre Drive*. It was widely discussed at a recent conference on starships. However the development of such a drive is far beyond the capabilities of Humanity for the foreseeable future – thousands of years. A major reason for its practical impossibility is that it requires the manipulation of masses of the size of Jupiter – utterly impossible. Can one imagine a "trick" that would make Alcubierre Drive, or any other gravitational drive, feasible. General Relativity gives a resounding No. General Relativity has been a subject of study for almost 100 years. *It is clear that the weakness of the gravity force, and thus the need for extremely large masses to have large gravitational effects, makes gravity driven starships impossible.*

1.2.3.3 Faster-Than-Light Quark-Gluon Drive Starships

There are many reasons why starships must go faster-than-light. Some of them have been discussed earlier. The primary reason can be seen in the history of world transportation. The trend of trade and migration between countries grew enormously as transportation became faster and vehicles became larger and safer. The transition from sailing ships to steamships to aircraft has had a large effect on shaping the present world.

As we reach out into the Solar System, and then to the stars, the speed and capacity of transportation must grow significantly.

There is only one apparent mechanism that can propel a starship from slow speed to speeds faster than the speed of light. This mechanism enables a vessel to circumvent the wall at the speed of light for real-valued speeds. The wall prevents particles accelerated by particle accelerators such as Fermilab or CERN from exceeding the speed of light no matter how much energy an accelerator gives them.

But if particles were accelerated to complex-valued[4] velocities they could avoid the wall and proceed to speeds much faster than the speed of light. Unfortunately accelerators cannot simply accelerate ordinary particles with real-valued velocities to achieve complex velocities. Accelerators can only create faster than light particles by collisions between atomic nuclei at very high energies.

Particle collisions at very high energies can create quarks and gluons with complex velocities. Lead, gold, and uranium ion collisions have created plasmas of quarks and gluons. Quarks and gluons have complex-valued velocities according to our derivation of The Standard Model of Elementary Particles. If we produce a stream of faster than light quarks and gluons to generate thrust[5] for a starship then they will give the starship a complex-valued velocity that will enable the starship to exceed, possibly vastly exceed, the speed of light.

Thus, in principle, we can make faster than light starships. There are major engineering feats that will be required to achieve this goal. But there is no "show-stopper" in this type of faster-than-light propulsion. We have the resources to do it with today's technology suitably upgraded. We do not need to juggle Jupiter size objects to obtain faster than light motion.

How do we "know" that quarks and gluons have complex-valued velocities? We have never seen an isolated example of either of these types of particles. The answer, which we will explore in much greater detail later, is a successful new form of The Standard Model that requires complex velocity quarks and gluons within it. This new form of The Standard Model embodies the many known features of the experimentally found Standard Model such as Parity Violation, the spectrum of particles, and The Standard Model's symmetries. One feature that has been verified in many experiments[6] (chapter 6) is faster

---

[4] A complex-valued number is a number that has a real-valued part and another part, the imaginary part, that consists of a real-valued number multiplied by i – the square root of minus one.

[5] This effort will require major engineering advances – primarily in increasing the flow of quarks and gluons. For example, extremely high power magnets will be required.

[6] There is one dissenting, unpublished experiment that found 5 neutrinos with speeds slightly below the speed of light. There are many experiments that find neutrinos move faster than the speed of light. See chapter 6.

than light neutrinos. This discovery, consistent with our theory, encourages the belief that quarks have complex velocities and thus our proposal for a starship engine based on quark ion thrust. Faster than light quarks do remain to be verified experimentally.

Building quark-gluon ion drive starships will enable us to achieve velocities of thousands of times the speed of light. At 5,000 times the speed of light a trip to the Centauri stars (about 4 light years away) is less than half a day. Accelerating to that speed and then decelerating will increase the trip time by months. Rapid travel to other stars in the galaxy and beyond would also become possible.

There is a potential problem in faster than light travel that we will address later. If a starship travels at a velocity much greater than the speed of light, then time on the starship "speeds up." One day of earth time would be 5,000 days of starship time at 5,000 times the speed of light. This problem can be fixed by putting the crew into suspended animation and having long lifetime starship equipment. The crew will sleep through an apparently long voyage but return to earth with little increase in their physiological age.

1.2.3.4 Nuclear Rockets for Travel in Other Star Systems

Because of the nature of faster-than-light travel[7] for distances of hundreds or thousands (or more) light years, nuclear rockets and chemical rockets for use at a distant destination must have an extremely long "shelf" lifetime – perhaps of the order of millions of years for especially long distance travel to other galaxies eventually. They would rest unused during a long starship time period and, upon arrival at a remote destination, they would be put into use. The implications of this requirement will be discussed in chapter 5 on long shelf life nuclear rockets.

---

[7] Later we will see that starship time progresses much faster than earth time on starships going much faster than the speed of light. A few years of earth time could be equivalent to thousands or millions of years on a starship. Thus the need for long-lived reactors and other components, and a corresponding need for suspended animation for starship occupants. Travel to nearby stars will have less stringent requirements for equipment lifetime and would not require suspended animation for starship occupants.

1.3 Space Program Stages: Prioritization Issues

The current approach of the space-faring nations is to use chemically powered rockets. Make bigger rockets to explore further in the solar system — particularly manned expeditions, and to build colonies in space, and on the moon, and possibly Mars.

Russia has also begun exploring an alternate path to space using nuclear powered rockets. Nuclear rockets have some decided advantages for longer trips in the solar system. For example a trip to Mars via chemical rockets takes about two years. A nuclear rocket could make the same voyage in about six months.

The United States has indicated preliminary interest in restarting a nuclear rocket program of the 1950's and 1960's.[8] Whether these ventures are brought to fruition or not they represent a fresh approach that would of great benefit for space programs.

With these comments as background we now outline a program that appears to be the most cost-effective and aggressive program for the next 100 years.

1.3.1 Near Term Program (Next 10 years)

The near term program proposed below would replace (actually defer) certain projects in favor of a long term cost-effective space effort.

1. Proceed to complete existing space projects that are close to completion financially. The NASA budget of $18 billion per year includes funding for these projects

2. Proposed projects for the near earth region using chemical rockets should be continued,

3. Projects for Mars and beyond should be suspended until efficient nuclear rockets become available. The outer planets

---

[8] A nuclear rocket program was started and then abandoned by NASA's predecessor, the National Advisory Committee for Aeronautics (NACA). For example, see "Steady Nuclear Combustion in Rockets" by E. Sänger, Astronautica Acta, I, Fasc. 2 (1955)

and moons will still be there when nuclear rockets can be used to explore them in a cost-effective manner.

4. The search for life in outer space should be suspended. While it has scientific value the primary interest in alien life is secondary to the much less expensive exploration of the deep oceans, deep mines and unusual areas such as Tibet, New Guinea, and the Amazon basin for new forms of life.

5. Studying these regions will make the eventual searches for alien life more revealing and informative.

6. The development of increasingly larger chemical rockets should be suspended and replaced by a powerful nuclear rocket development program. The nuclear rocket can be constructed in pieces on earth and shot into space in pieces for assembly in an earth orbit at a distance of perhaps 10,000 miles.

7. The need to launch large quantities of materials into earth orbit requires launch methods other than rockets. Rocket propulsion is extremely expensive and should only be used for fragile cargo and people. One possible cost-effective method might be multi-stage space guns. A hint of the potential of this method is the ability of a large gun to shoot a one kilogram payload (shell) about 100 miles into space (called Near Space) using only one kilogram of propellant. Compare this cost to the cost in rocket fuel to send a one kilogram payload into Near Space. Yes, there are technical issues and subtle cost issues such as refurbishing the gun barrel after some shots. But the cost savings of a successful space gun would be substantial and would be very helpful in constructing a nuclear rocket in space. NASA had a space gun study program in the 1960's that was terminated. However with the development of new alloys and ceramics in the past 50 years, a reinvestigation of space guns is reasonable. We

propose a modified multi-stage space gun in chapter 3 that should be more efficient and less costly than the space gun studies of the 1960's.

1.3.2 Long Term Space Program (50 years?)

In this section we will consider a Long Term Space Program directed towards travel to the stars. This program can run concurrently with the Near Term Program, which is solar system oriented. The initial expenditures for the Long Term Program are not large. Consequently the reprioritized NASA program could cover the costs of both programs with little additional funding

To those who view the current budget deficit of such great concern that NASA's budget should be cut – not increased – we can only say that the costs of delay in executing these programs far outweigh the small savings (compare to the trillion dollar deficit level) that would accrue from wielding the budget axe on NASA's rather small 18 billion dollar budget.

So we propose a program directed towards the stars. We will divide the starship program into two parts: a starship program for travel to nearby stars within a 20 – 30 light year radius of earth – the Short Distance Starship Program; and a Long Distance Starship Program initially designed to travel at very high speeds for thousands of light years initially (in our galaxy), and then extended to create starships designed for truly long range travel to other galaxies.

The optimal cost-effective approach to starships is:

1.3.2.1 "Short" Distance Starship Program

This starship program is for travel to the nearby approximately 100+ stars within a 20 – 30 light year radius of the earth. Alpha Centauri, only about 4 light years away is now known to have one earth-like planet. The three Centauri stars are likely to have more earth-like planets – some of which may be in the habitable regions[9] around these stars. Earth-like planets are also very likely to exist circling other stars in

---

[9] At such a distance from a star that liquid water might exist on the planet's surface, and temperatures would be comparable to those of earth.

the "neighborhood" of earth. Tau Ceti which is 11.9 light years away appears to have at least two earthlike planets circling it in its habitable zone – one planet is four times the size of earth and has a 6 month year.

The program must begin with the development of a starship engine capable of speeds up to 10c (10 times the speed of light). Then reaching a star 30 light years away could take a few years of earth time taking account of the time required to accelerate and decelerate the starship. As we will see the starship crew would experience a trip time of perhaps thirty years or so. Ideally the crew could be put into suspended animation so that they would not age very much. Thus the Short Range Program would contain:

1. The design and development of a starship engine that would be capable of speeds of 10c. The only possible approach to a faster than light starship engine within the range of humanity's resources is the Faster-Than-Light Quark-Gluon Drive Starships discussed briefly earlier and discussed in more detail later in this book. This approach has to be proven first. It is based on a new theory of the Standard Model of Particle Physics developed in numerous books by this author. This theory has the exciting feature of explaining/deriving the Standard Model which had previously been considered as a provisional model by most particle physicists. The design and development effort can take place on earth. A reasonable cost estimate for this effort is $30 billion dollars spread over fifteen years - $2 billion per year. (This estimate is based on the cost of the CERN LHC accelerator since the scale of the starship engine appears to be similar. The additional monies reflect the redesigns and redevelopments that are to be expected in creating a new engine.)

2. Concomitant with this R&D effort other aspects of the project should be developed – not only for use in the project but also for their own intrinsic value for earth-based technology. These efforts include the development of long-lived chip technology so that we may have long life

computers and other electronic equipment, the development of nuclear space ships, and the development of suspended animation for people. We will address these topics, and additional efforts such as "instantaneous quantum communication" in more detail later.

3. The development and construction of a nuclear space ship. A working ship should be constructed in orbit and used in voyages in the solar system. The experience gained in the use of the ship will lead to further improvements in nuclear space design – just as we have seen in the history of automobiles and airplanes.

4. Assuming a successful starship engine prototype is constructed on earth, a starship prototype should then be built in space and tested in trips around the solar system. Again the experience gained in actual use should lead to design improvements that can be implemented in the construction of the Short Distance Starship for travel to nearby stars.

1.3.2.2 "Long" Distance Starships

Long distance starship – starships that can travel up to 100,000 light years from earth - will be more or less the same structurally as short range starships with some important differences. The primary differences will be a much larger size necessitated by the need for vastly more fuel, and engines and other components such as computers with extremely long lifetimes of the order of thousands to millions of years of starship time,[10] and a form of suspended animation that could keep the crew in suspended animation for up to millions of starship years.

The development of these capabilities will take decades and will involve R&D in materials science, physics, and biomedicine. The cost of this R&D will be many billion dollars spread over a time period of the order of forty years plus or minus. The combined yearly R&D costs would probably amount to hundreds of millions of dollars. The program should independently fund the long lifetime materials science and

---

[10] Although the elapsed equivalent earth trip time would typically amount to a few earth years.

component engineering R&D, and the biomedical research into long term suspended animation. Another major area of development will be the refinement of the starship engine configuration to have a long lifetime, and increased efficiency in power utilization for starship thrust – "a longer life and better mileage."

For each of these long range starship requirements prototypes and/or extensive testing would be required. For example simulating a computers performance over a million year period – particularly with respect to chip deterioration would be required – as would simulations of million year performance of starship machinery and suspended animation. In this effort we would be following general procedures developed by individuals to age art/archaeological forgeries for sale to unsuspecting buyers – but in our case for a worthy cause.

The end result would be the construction of a long range starship in space and trial expeditions to stars perhaps within a thousand light year radius. As our experience, confidence and capabilities increase, longer trips including trips to nearby galaxies such as the Magellanic[11] Clouds should be undertaken.

1.4 Technology Spinoffs Benefits

The Starship Programs – like the US Moon program of the 1960's – gives us a national goal of great symbolic importance in an age where we face a decline in earth's environment. It also provides a worthy, positive program that will stimulate economic and technological growth. As General Patton once said in World War II, "It's better than shoveling ... in Louisiana."

Some of the spinoff benefits we envision for the starship programs are listed below. We also anticipate "unexpected" spinoff benefits from these R&D efforts as has happened so many times in previous R&D projects. Some of the spinoff benefits of the starship programs, and related programs, are:

---

[11] Magellan – aptly named after a great explorer.

1. A major move into space with the mass transport of materials via multi-stage space guns. Moon and Mars colonies become more affordable.

2. New developments in nuclear reactor technology for power production on earth.

3. The development of improved electronic and computer devices.

4. Major medical applications of suspended animation for patients on earth.

5. Advances in magnet technology.

6. Possibly the development of fusion power through experimentation in starship engine R&D.

7. The development of new materials for long lifetime machinery and computers.

8. Possible development of new instantaneous, secure, communication methods based on quantum effects.

All in all the envisioned starship programs carried through to completion will provide major additional benefits and advances for Mankind.

1.5 Benefits of Travel to the Stars

If we succeed and travel to the stars there will be significant benefits besides the thrill of discovery. Some of these potential benefits are:

1. The possible discovery of metal and other ore bodies that could be extracted and transported to earth. Examples include gold, rare earth metals, diamonds, and so on.

2. The discovery of new forms of materials created under peculiar conditions. An example of a new form of compound only recently found on earth is a buckyball. The stars may have more new, unusual compounds. (These materials would consist of compounds composed of the atoms known to earth science – but in combinations/configurations that had not been discovered on earth. There is also the somewhat remote possibility of finding atoms from beyond the periodic table of chemical elements in some "island of stability" that earth laboratories have not found as yet.)

3. If life exists "out there" as it most likely does, then we will find new types of plants and animals. As discoveries in the Amazon and earlier in the Americas show plants found on planets circling other stars may important food and biomedical benefits. For example potatoes and corn were found in the Americas and became important food sources in Europe. Many biomedical wonder drugs were extracted from plants such as penicillin.

4. Exploring the structure of planets circling other stars can provide us with information on planetary dynamics that can help us better understand earth's dynamics including the nature and prediction of earthquakes and volcanos.

5. There are many earth-like stars. Some of them may harbor intelligent life. If so, then the possibility of information exchanges on technology, knowledge they may have of other star systems in the galaxy, culture, and philosophical/religious thought would be of great value.

1.6 Negatives of Travel to the Stars

There are a number of dangers and negatives associated with travel to the stars. Some of the major issues are:

1. The starship may malfunction for many possible reasons causing the death of the crew. Some reasons are: engine failure,

breakdown of suspended animation, the impact of meteorites of various sizes including large quantities of high speed space dust, and so on.

2. Encounters with poisonous plants or dangerous animals on remote planets leading to crew fatalities.

3. Failure of shuttles used to land on planets.

4. Psychological issues associated with travel time that may arise.

5. Encounter with a hostile, intelligent life and consequent strife.

6. Acquisition of alien knowledge/philosophy that would lead to socio-psychological problems in the crew.

These negatives are significant. But they have been faced countless times since the emergence of the human race and so should not hold back the expansion of humanity to the stars.

2. A Crowded, Declining Earth and an Inhospitable Solar System

One wishes to begin a book on a hopeful note if possible. But the present circumstances of the earth give little grounds for hope on earth, or on other planets and moons in this solar system. In this chapter we outline the serious problems facing our planet and point out that currently there are no feasible means of large scale colonization of other bodies in this solar system.

2.1 Earth in Decline

The decline in the earth's environment is indisputable. The hope that the decline can be reversed appears to be unrealistic without drastic steps that are not acceptable to a humane, democratic civilization. A list of the ongoing declines is below. The issues are well known.

1. The Decline in the Earth's Arable Land
    About 10% of the earth's surface is suitable for cultivation. 25% of that arable land is hopelessly degraded and not suitable for cultivation due to pollution, growing deserts, and soil exhaustion. While one can hope for new miracle crops to feed earth's growing population it is not likely. If such miracle crops were developed then the fecundity of the soil in which these crops were grown would decline. One could argue that exhausted soil could be "renewed" by fertilization. But fertilization has been found to introduce additional problems. For example, recycling sewage into fertilizer leads to heavy metals being put into soil as well as bacteria and viruses. Global warming also appears likely to reduce the amount of arable land.

2. The Decline in Clean Fresh Water

The world is experiencing a decline in the quantity of clean, fresh water per capita due to pollution, global warming, draining of non-renewable underground aquifers and population growth. As a result clean water for the population and industry is becoming an increasingly large problem.

3. The Decline in the Oceans

The world's oceans are becoming increasingly polluted and more acidic. One major problem, exacerbated by the growth in the price of gold, is the increasing mercury content in the oceans. As a result an important food resource – consumable fish – are becoming health hazards. Many parts of the world rely on fish for a major part of the protein in their diets. The decline in the quality of ocean water will have a direct negative effect on the world food supply. As it is, many types of fish are being depleted by overfishing.

4. Overpopulation and Declining Standard of Living

The growth in the world's population has made the achievement of higher standards of living impossible. The world is more or less in a zero sum position economically. The growth in the living standard of one country or region now comes at the expense of a decline in the living standard of another country or region. A clear example of this phenomenon is the transfer of wealth from the United States to China and India. The living standard of the United States is in decline (although masked by massive borrowing.) The living standards of China and India are rising - but slowly due to the large size of their populations.

2.2 Planets and Moons of Our Solar System

Our solar system has many planets, moons and asteroids with interesting features that make them well worthy of study and exploration. Active exploration programs are underway and are yielding many novel results.

One major effort is focused a search for extraterrestrial life. Given the nature of the planets and moons, only simple forms of life such as virus-like or perhaps bacteria-like creatures are likely to be found. The most important result of such a find would be its effect on religious and philosophical thought. This author is not aware of any absolute statement by a major religion that life only exists on earth.

The scientific import of finding extraterrestrial life is difficult to prognosticate. Chances are that it will be carbon-based and structurally similar to corresponding earth life. This hypothesis is based on the evident tendency of Nature to imitate successful designs. One possible practical benefit of finding extraterrestrial life is finding new biomedical drugs that could be useful for Mankind. Another benefit is the possibility of expanding our understanding of the nature of life. Lastly, one must remember that extraterrestrial life could be harmful to Mankind. So a suitable quarantine procedure should be in place before samples are brought to earth.

Beyond the search for scientific knowledge the possibility exists for the creation of colonies in space and for eventual mass migration into space. When considering these possibilities the environment of planets and moons must be considered. Briefly put, the planets and moons are not viable locations for large colonies without massive expenditures. These large expenditures do not seem to have a reasonable justification.

Except possibly for Mars it does not seem possible to change a planet or moon to make it hospitable without massive expenditures. In the case of Mars it might be possible to divert a watery asteroid(s) to collide with Mars to create a water-laden carbon dioxide atmosphere that would warm the planet and support plant life that would produce a level of oxygen that would be equivalent to mountainous regions of earth such as Mexico City or Tibet. Then it becomes possible to have major colonization at a reasonable cost.

3. Multi-Stage Space Guns For Cost-Effective Cargo Shipment to Near Space

In the 1960's NASA investigated alternative approaches to chemical rockets due to the high cost of sending cargo and crews into space using chemical rockets. In this chapter we will consider another approach to sending material into space through the use of "space guns" of the type proposed by Jules Verne and others. In subsequent chapters we will consider nuclear rockets – also studied by NASA in the 1950's and 1960's. Due to the dangers of a nuclear accident it is not reasonable to use nuclear rockets to lift material into space from the earth.

So it appears that chemical rockets and space guns would be the preferred transport vehicles for transportation from earth to space although there are other proposals being considered. This author does not believe the other proposals, possibly excepting rail guns, are practical. In the studies of space guns in the 1960's it was noted that the gun barrels were quickly worn. Other issues also cropped up. However with advances in metallurgy and ceramics it is possible that space guns made with modern materials may be a cost-effective alternative to rockets. In this chapter we will discuss a new multi-stage space gun design that might make the mass transport of cargo to space more economical.

3.1 Single-Stage Space Guns

The practicality of space guns became apparent in World War I. German big guns (the approximately 100 foot long Big Bertha and a larger gun) that bombarded Paris from a distance of 80+ miles sent their 100+ lb shells as high as 80 – 90 miles above the earth to the very edge of space –now known as Near Space. The shells had a speed of 1 mile per second as they emerged from the gun's barrel.

The power of the German big guns was exceeded by the guns developed in the NASA HARP program led by Jerry Bull in the 1960's.[12] In Project HARP a 16 inch Navy gun was used to send a 180 kg shell to a height of 180 km – Near Space.

Space guns are interesting in the light of the fact that altitudes of about 100 miles are viewed as "Near Space." Today many space satellites circle the earth at 200+ mile altitudes in "low earth orbit". Space guns can put objects into Near Space. The price to shoot one kilogram (2.2 pounds) at a muzzle speed of 1.6 km/sec is about one kilogram of the best chemical propellant. In comparison a rocket would use many times more propellant to lift a kilogram up eighty miles. Thus space-guns potentially are much more cost-effective.

A single or multi-stage rocket, with or without boosters, uses over 99% of its weight as fuel to send a payload into space. The fuel is, for the most part, used to propel the rocket itself off the ground at an ever-increasing speed into space.

In contrast, the propellant for a *single-stage* gun[13] propels the projectile and a fraction of the propellant between the point of burn of the powder charge and the projectile. Thus a sufficiently large gun could efficiently put a payload up eighty miles into Near Space because the propellant propels the payload, and not fuel or a rocket casing. If the space gun is scaled up to send 500 kg payloads into space then we have an effective mechanism to send large amounts of material 80 miles into space in 500 kg chunks using about 500 kg of propellant per payload.[14]

When the payload reaches 100+ miles or so then there are two possibilities. The payload might contain a small rocket to put it into a higher earth orbit or to send it to a space station. Or there could be a

---

[12] I am grateful to Dr. Mitat A. Birkan, Program Manager, Space Propulsion and Power, AFOSR/NA for providing this information as well as other details discussed later on chemical space guns.

[13] By single-stage gun we mean a gun of the type of Big Bertha. We discuss multi-stage space guns later in this chapter.

[14] Dr. Mitat points out that the amount of propellant increases rapidly with muzzle velocity and that at over 2 km/s the required propellant mass is more than three times as large as the payload.

"scooper" vehicle circling the earth that could scoop up the payload and deliver it to a space station at a higher altitude.[15]

The fabrication of space guns is well within our technology. The major drawback to space guns is the deterioration of the wall of the gun barrel with repeated use. However, the walls can be resurfaced, presumably, at a much lower cost than replacing a throwaway rocket or refurbishing a used space shuttle. Another issue is the design and construction of scooper vehicles. This does not appear to be a significant problem since constantly circling vehicles using atmospheric bounce were designed in the 1950's by NASA although never built.

With the new stronger alloys and ceramics that have been developed in the past 50 years, gun barrel walls could be made of greater strength and endurance reducing this problem significantly.

Thus single-stage space guns deserve renewed consideration for cargo transport. In the next section we discuss a new design for space guns – multi-stage space guns that could reduce propellant requirements significantly making them more cost-effective than a single stage space gun.

3.2 Multi-Stage Space Guns

The cost of sending cargo and personnel into space is the overwhelmingly dominant factor hindering Mankind's move into space. In the preceding section we considered single-stage space guns – guns with a simple barrel containing a projectile and propellant. In this section we consider some possible designs for space guns that might significantly reduce the propellant requirements for projectile propulsion to Near Space.

The single-stage space guns considered in the 1960's consisted of a simple barrel containing a shell with propellant beneath it (Fig. 3.1).

---

[15] The concept of a vehicle constantly circling the earth at a range of altitudes and "bouncing" off the earth's atmosphere to reach higher altitudes was developed by German scientists in World War II and studied by American scientists after the war. This type of vehicle could be used as a scooper vehicle for payloads shot into Near Space. Recently there has been a revival of interest in this concept.

3.2.1 Basic Multi-Stage Space Gun

The basic multi-stage space gun in Fig. 3.2 has a barrel with chambers loaded with propellant. The propellant chambers are placed at ever decreasing distances as one goes from the gun bottom to the top. As the cargo shell accelerates up the barrel the chambers are carefully exploded in order to provide maximum thrust to the cargo module. The time order of the chamber explosions can be accurately set electronically. In addition to the explosion chambers along the barrel there will be a propellant charge under the cargo module to initiate liftoff. The combined effect of bottom propellant charge and the propellant charges in the barrel chambers should enable the cargo module to attain the same velocity as a single-stage gun with much more propellant. (An additional benefit is the multi-stage gun barrel should have less wear relative to the single-stage gun barrel since it builds the velocity of the cargo more slowly thus lowering the gun barrel friction – especially near the base of the gun.)

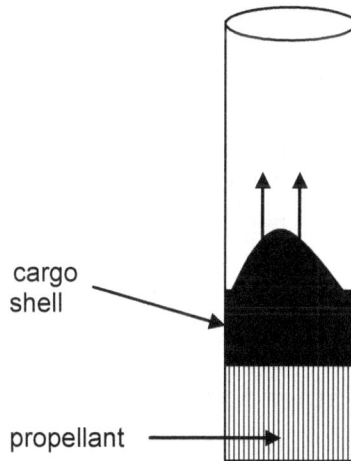

cargo
shell

propellant

Figure 3.1 Conventional Single-Stage gun before ignition.

If we assume that it takes a propellant mass M for a single-stage space gun to attain a certain exit speed, and the multi-stage gun takes the sum m of the bottom propellant and the propellant in the barrel chambers, then we save propellant if m is less than M. We achieve the saving primarily by not using propellant to propel the propellant above it upon ignition as happens in a single-stage space gun.

The multi-stage space gun is analogous to a multi-stage rocket. Propellant is used to propel the cargo. In a multi-stage rocket, sections of the rocket drop away after their fuel empties. In a multi-stage space gun, propellant is not used to propel the yet unburned fuel above it, as is the case for a single-stage gun.

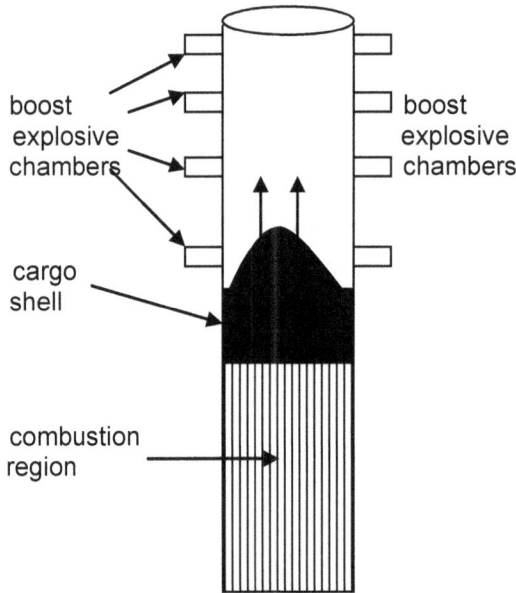

Figure 3.2 Four stage space gun after ignition of propellant below cargo module.

3.2.2 Enhanced Multi-Stage Space Gun

Reducing fuel consumption is a significant space gun goal. One method of achieving this goal is to have a piston beneath the propellant in the multi-stage gun as depicted in Fig. 3.3. The piston would move upward in step with the ascending cargo module helping to maintain the upward pressure on the module. With this addition it may be possible to achieve further cost benefits in propellant – even taking account of the propellant (energy) required to push the piston. The motive force for the piston can be a chemical explosion, a mechanical spring mechanism, or a powerful electromagnetic field.

3.3 Other Measures to Minimize Space Gun Costs

There are a number of other measures that could lead to further reductions in the cost of space gun shots and gun maintenance as well as enabling more rapid shooting of space guns. Some possibilities that come to mind are:

1. Since gun shots wear the barrel it might be desirable to have an inner removable sleeve that could be replaced after some number of space gun shots.
2. The explosion chambers might be set up in such a way as to resemble ordinary gun magazines to expedite space gun shots.
3. The loading of propellant and cargo modules could be done from the top of the gun.
4. The bottom and barrel of the gun could be rapidly cooled between shots to increase the shooting rate.

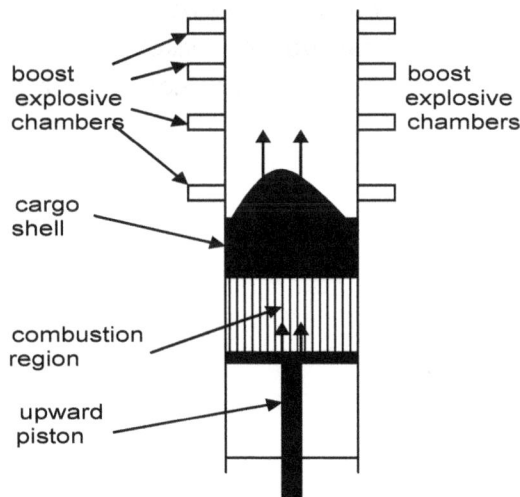

Figure 3.3 Multistage space gun with explosive chambers along the barrel to boost the payload cargo "shell." A fast upward moving piston keeps the combustion region small.

The space gun goal gives rapid transport of large cargoes to Near Space in the most cost-effective manner. If successful, it would enable rapid, low-cost construction of space stations for communications and scientific R&D. It would also enable us to build nuclear and other rockets in space for travel to other planets and moons.

3.4 Moving Massive Amounts of Cargo Into Space

The multi-stage space gun concept that we have developed supports the possibility of moving massive quantities of cargo into space including satellites, space station components, fuel, and space ship components for conventional and nuclear rockets. It could also place cargo into orbit which can then be sent to the moon to construct a moon base.

The key to massive cargo transport is to have an array of sufficiently separated space guns that might each perhaps be capable of shooting 500 kg to space once or twice per week. The time between shots can be used for maintenance, propellant loading and cargo loading. A hundred such space-guns could send 100 metric tons of material into Near Space (100 miles up) if each gun was shot twice per week – at the cost of perhaps 150 metric tons of propellant – far less than the fuel required for transport by rocket.

This capability opens the door to cost-effective space travel and colonization.

4. Nuclear Rockets: The Path to the Planets

While the use of nuclear rockets to transport cargo from the earth into space is risky due to the possibility of malfunctions that might lead to a Chernobyl-scale disaster, nuclear rockets are the fastest and most economical means of transportation in the region beyond the dominance of the earth's gravity field to the moon and the planets and their moons.

There are four known basic types of nuclear rockets:

I.   Nuclear rockets where nuclear reactions take place in a combustion chamber generating rocket thrust directly.[16]

II. Nuclear rockets that generate electricity which is used to ionize and accelerate particles to power an ion drive for thrust.

III. Nuclear rockets that directly heat inert fuel such as hydrogen to use as thrust. (Nuclear Thermal Rockets, also called Bimodal Nuclear Thermal Rockets)[17]

IV. Nuclear rockets where multi-kiloton nuclear explosions "push" the rocket by exerting pressure on an "umbrella" or similar structure on the rocket.

The basic requirements for a nuclear rocket to range the solar system are:

1. Sufficient speed and fuel capacity to range from Mercury to Charon. (The initial nuclear rocket might have a restricted range from Venus to Jupiter.)

---

[16] A nuclear rocket of type I was considered by E. Sänger, Astronautica Acta, I, Fasc. 2 (1955).
[17] The NERVA project discussed later was a type III nuclear rocket.

2. A service lifetime of approximately 25 years.
3. Adequate protection for the crew from rocket and space radiation during long flights.
4. Safety.

4.1 History of Nuclear Rocket R&D

There have been numerous proposals and several R&D programs to develop nuclear rockets. At least one of these R&D efforts, the NERVA program of the 1960's, reached the test stage and a proposal for a 1978 manned landing on Mars was put forward. Federal budgetary considerations caused the cancellation of the NERVA Mars project. In this section we discuss a number of nuclear rocket projects. We will consider the NERVA program and a new Russian nuclear rocket program in section 4.2.

Some of the noteworthy nuclear rocket studies and projects were:

    a. Project Orion – based on "nuclear pulse propulsion."
    b. Project Longshot – a US Naval Academy and NASA also based on "nuclear pulse propulsion."
    c. Project Daedalus – a British Interplanetary fusion rocket study.
    d. NERVA – A NASA nuclear thermal rocket program
    e. Prometheus – a NASA project to develop nuclear propulsion for long-duration spaceflight (2003).
    f. The Bussard ramjet
    g. A Fission-fragment rocket
    h. A Fission sail
    i. A Gas core reactor rocket
    j. A Nuclear salt-water rocket
    k. Radioisotope rocket
    l. A Nuclear photonic rocket

4.1.1 Bimodal Nuclear Thermal Rockets

Bimodal nuclear thermal rockets appear to be the current frontrunner in nuclear rocket development. They use nuclear fission reactions similar to those at most nuclear power plants. The energy generated by a nuclear reactor heats liquid hydrogen propellant, which provides the thrust for nuclear rockets. One advantage of this type of nuclear rocket is they are radiation free at launch time. These nuclear rockets would be constructed in space and used to travel to other planets and moons. Nuclear thermal rockets would have major performance advantages over chemical rockets. They appear to be at least twice as efficient as chemical rockets.

4.2 NERVA Program Overview

NERVA is a bimodal nuclear thermal rocket project that achieved major successes in the 1960's but was cancelled for budgetary reasons by the Nixon administration. NERVA (an acronym for Nuclear Engine for Rocket Vehicle Application), was an effort of the U.S. Atomic Energy Commission and NASA until ended in 1972. (See Dewar, 2008).

NERVA showed that nuclear thermal rockets could safely engage in space exploration. In 1968 the NRX/XE NERVA engine was shown to meet all the requirements for a manned Mars mission. NERVA engines were built and tested as much as feasible on earth. The NERVA program was cancelled by the Nixon Administration for budgetary reasons and to avoid a growing space contest for travel to other planets. The specifications for the NERVA rocket were:

- Diameter: 10.55 metres (34.6 ft)
- Length: 43.69 metres (143.3 ft)
- Mass empty: 34,019 kilograms (75,000 lb)
- Mass full: 178,321 kilograms (393,130 lb)
- Thrust (vacuum): 333.6 kN (75,000 lbf)
- ISP (vacuum): 850 s (8.09 kN·s/kg)
- ISP (sea level): 380 s (3.73 kN·s/kg)
- Burn Time: 1,200 s
- Propellants: LH_2
- Engines: 1 Nerva-2

The NERVA testing program spent 17 hours in operation – 6 of which were above 1700° Centigrade. NASA had plans to send a NERVA class ship to Mars by 1978 and to establish a permanent lunar base by 1981. The runtime of a NERVA rocket was only limited by the amount of liquid hydrogen propellant. Unfortunately the NERVA program (which ended in the Pewee rockets) was cancelled by Congress in 1978 despite its success.

Recently in 2010 design studies of nuclear rockets, NERVA-Derivative Rockets or NDRs, based on Pewee designs took place. With friendly Russian competition in mind it appears that the United States will not only design and test earth-based prototypes but also construct a nuclear prototype in space as a first step to nuclear rocket travel to other planets.

4.3 Proposal for Nuclear Rocket Mars Missions vs. Ultra-large Chemical Rockets

A proposal is currently being considered (December, 2012) to build an enormous rocket at a cost of roughly $600 million that is capable of a manned flight to Mars in 2028. This rocket would be an expanded version of the large chemical rockets that NASA has been using for the past 50 years. The trip to Mars in this rocket would take approximately 2 years.

An alternate approach, that should be seriously considered, is a restart of the NERVA program taking advantage of new technology and materials that have been developed in the past thirty years. The NERVA program tests in 1978, and successor program tests afterwards, were extremely successful and a NEW-NERVA nuclear rocket project would be cost competitive with the giant chemical rocket project, enable much shorter trip times of the order of weeks – not years, and would be reusable for manned trips to other planets, asteroids and moons.

In short, the NEW-NERVA project would be a cost-effective approach to support many trips in the Solar System. The project could be enhanced by using not only hydrogen fuel but also methane and other liquids found on the moons of Jupiter and Saturn as fuel.[18] Thus

---

[18] Refueling should be flexible. The ship should be able to extract fuel from moons in solid, liquid and gaseous form. The knowledge gained in the oil industry in extracting oil in a variety forms on land and in oceans as well as in regions ranging from the Arctic to the tropics should be of great help in developing the capabilities for fuel extraction from bodies in space.

the range of NEW-NERVA ships could be vastly extended by refueling on Jupiter's and Saturn's moons.

4.4 New Russian Nuclear Rocket Program

The value of developing nuclear rockets has been recognized by the Russian Federal Space Agency in its new NPS Development Project. Anatolij Perminov, the managing director of the Russian Space Agency has announced that it is developing a long-range, bimodal nuclear thermal rocket. Its design should be completed by 2012. It will take 9 more years for the construction of the rocket in space. The cost is estimated to be about $600 million dollars.

A Russian nuclear rocket would consist of a nuclear reactor(s) and a set of ion engines using hydrogen propellant. Perminov stated the rocket could be used for a manned mission to Mars with a 30 day stay on Mars. The trip to Mars would take 6 weeks of steady propulsion - instead of an 8 month journey using a chemical rocket. Perminov bases his estimate on a nuclear rocket with a thrust 300 times greater than a chemical rocket. Work on the Russian rocket is apparently underway. The next Sputnik?

4.5 Pulsed Pellet Micro-Nuclear Explosion Rocket Drives

While the NERVA project nuclear engine looks like a good initial start on a nuclear rocket, there are a number of developments in the past forty years that might make it possible to develop a more efficient nuclear rocket powered by pulses of micro-nuclear explosions of fissile pellets.[19]

Until now nuclear rockets, in our discussions, were conceived as one of the first three types listed at the beginning of this chapter. Type IV nuclear rockets have also been proposed by a variety of individuals and organizations including Project Orion, Project Daedalus of the British Interplanetary Society, and Project Longshot of NASA and the US Naval Academy. See Bussard (1965) for additional proposals.

---

[19] A fissile pellet might be a sphere of U-235 of the size of a pinhead and have a mass of the order of a gram.

We now consider the possibility of using micro-nuclear explosions to propel a rocket. We propose to use an assembly of many "small" thrusters to propel a rocket with each thruster generating thrust by small nuclear explosions. One normally expects that a large mass of U-235 or some other fissile (fissionable) atom is required for a nuclear explosion or nuclear reactor. Nuclear explosions typically require 10+ pounds fissile matter. However, under the right conditions of temperature and pressure much smaller amounts (pellets) of fissile matter can be supercritical.[20] In a nuclear bomb there are two methods to initiate a nuclear explosion: 1) collide two chunks of sub-critical fissile matter together at high speed to create a supercritical compacted supercritical mass that then explodes, or 2) compress a sub-critical mass to much higher density (pressure and temperature) to create a supercritical mass that then explodes. Both methods typically use high explosives to initiate the supercritical state.

4.5.1 Pulsed Pellet Micro-Nuclear Explosion

A Pulsed Pellet Micro-Nuclear Explosion Rocket Drive takes sub-critical pellets of fissile matter and accelerates them through one of a number of means. Then it collides them in pairs at extremely high energy. The result is a mass with the temperature and pressure (and high density) to make it supercritical with an ensuing nuclear explosion. The nuclear explosion can be directed by strong magnetic fields to generate a thrust of high speed of the order of 1% to 3% of the speed of light.

The process can be viewed as a series of stages:[21]

1. Feed pellets into two accelerators (accelerating rings or linear accelerators) to bring them up to a speed which, upon head on collision, creates a supercritical, fissile globule. A plasma wakefield accelerator (discussed later) might be developable as the accelerator mechanism.

---

[20] In a state that can spontaneously sustain a fission chain reaction.
[21] See Fig. 4.1 for a schematic diagram of the configuration.

2. The pellets enter a magnetic bottle of the type being investigated in fusion R&D with an "opening" towards the rear from which the thrust exits after the colliding pellets undergo fission. The bottle might contain a gaseous vapor that would enhance the fission process. Some implosion nuclear bombs do use such a catalyst to accelerate fission.
3. At the time of a pellet collision an array of laser (or electron) beams compresses the pellets enhancing the supercritical state of the colliding pellets.
4. The magnetic bottle then directs the fissioned pellets' fragments towards the rear generating the thrust.

An array of pulsed pellet nuclear modules can propel a rocket. The number of modules would be determined by the power requirements of the rocket. A simple rocket design could be based on an array of Pulsed Pellet Micro-Nuclear Explosion modules at the tail of the rocket.

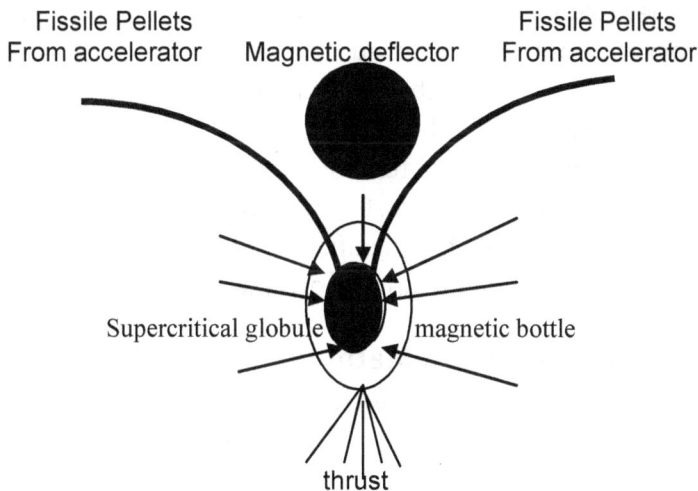

Figure 4.1 Schematic of pulsed pellet nuclear rocket engine module. Two beams of compressed fissile pellets are collided to create a supercritical globule that undergoes fission. The beams enter a powerful magnetic bottle that allows the fission fragments

to exit in one direction providing the rocket thrust. This design might be characterized as quasi-confined inertial compression fission. The arrows symbolize laser (or electron) beams that provide compression of the pellets and collision region to achieve fission. The rocket engine undergoes repetitions of these nuclear explosions to generate a continuing thrust. Magnets are used to direct the incoming beams and outgoing fission fragments thrust. Many units of this type can appear on a nuclear rocket to combine to produce a powerful thrust.

The most important question for Pulsed Pellet Micro-Nuclear Explosion Rockets is whether the power generated by the fission of colliding pellets is significantly more than the power required for the acceleration of the pellets, their deflection by magnets to collide, their compression by laser/electron beams to a supercritical state, and the direction of the collision fragments by magnets to create a thrust exhaust stream.[22] This process appears to be efficient. Bu this question can only be answered by detailed R&D studies. If this form of nuclear engine is less efficient than thermal nuclear rockets then thermal nuclear rockets would be preferred.

The resemblance of this design to ongoing attempts for fusion energy devices should not have escaped the reader's attention. This approximate similarity creates an interesting synergy between information learned in this effort and in fusion energy research. Thus this R&D effort potentially has significant side benefits for fusion energy research. It might also lead to small portable nuclear power reactors.

4.5.2 Pellet Acceleration Methods

There are two standard ways to accelerate particles or pellets to very high energies of the order of a GeV or higher. Circular accelerators accelerate pellets in circular rings to higher and higher energies. At the desired energy level pellets are siphoned out of the ring with magnets and sent to the collision point. In the case of pellets they can be made

---

[22] It is possible laser/electron compression – called implosion – may not be necessary if the collision energy is sufficiently large. Implosion approaches were used to achieve fission in low yield nuclear weapons – powerful explosives compressed the fissile matter.

charged by stripping electrons from atoms in the pellets and then inserting them in a ring. A second ring performs the same steps but accelerates the pellets in the opposite direction so that pellets from each ring can collide at high energy.

An alternate approach is to use linear accelerators that accelerate pellets in a straight line.

4.5.3 Pulsed Pellet Fission-Fusion Explosion Rocket Drives

It is possible that the Pulsed Pellet Micro-Nuclear Explosion Rocket modules could be upgraded to a more powerful and efficient Pulsed Pellet Micro-Nuclear-Fusion Explosion Rocket.[23] The design that we described for the Pulsed Pellet Micro-Nuclear Explosion Modules may be adaptable to a combined Micro-Nuclear-Fusion module. The key new factor is to collide a stream of pellets at very high energy while simultaneously compressing them with laser or electron beams. The pellets would be charged with an outer layer of fissile material such as U-235 with a center composed of a layer of fissile material such as lithium deuteride enclosing a center composed of a deuterium/tritium mixture. The explosion of the fissile material would cause an implosion that would ignite the fusion material within – a form of Inertial Confinement Fusion.

There are a number of major R&D issues facing this mechanism. Among them are the geometry of the pellets, and the layering of the pellets so that premature explosion of the layers doesn't abort the mechanism.[24]

If this mechanism proves successful then Nuclear-Fusion rockets would substantially increase our space capabilities in a cost-effective way.

---

[23] The British Interplanetary Society designed a fusion starship in 1973-78 in a project called Daedalus. Their mechanism was based on igniting small pellets of fusionable fuel using a deuterium/tritium trigger at the center. A pellet is hit from all sides by lasers or electron beams compressing the pellet to the point of fusion (Inertial Confinement Fusion). The resulting heated plasma forms the thrust. The difficulties of initiating a fusion reaction for energy production currently facing energy projects suggest this approach will not work as Daedalus designed it.

[24] Another issue is pellet lifetime. Leakage of deuterium and tritium limits the lifetime of the fuel.

4.6 Plasma Wakefield Accelerators

While circular accelerators and linear accelerators have been used to accelerate particles for many years, a new approach to accelerators, plasma wakefield accelerators, appears a promising way to achieve acceleration to high energies in a more compact and cost effective way.

A plasma wakefield accelerator has a beam cavity that contains a plasma. The particles or pellets being accelerated have an accompanying pulse of protons, electrons or laser light. The charged particles in the plasma are forced by the pulse to cluster around the particle/pellet and to move behind the particle/pellet thereby accelerating the particle/pellet. A plasma wakefield can create steep energy gradients (increases). Recent experiments have obtained gradients as much as 200 GeV per meter over short distances of the order of millimeters – much more (by a factor of 10 or more) over linear accelerator gradients.

Thus plasma wakefield acceleration can potentially accelerate pellets to high energies using smaller, less costly (in size and mass) equipment. Plasma wakefield acceleration is currently an active area of research and could become the means of accelerating pellets in nuclear and/or nuclear/fusion rockets.

4.7 Suggested Initial Flights of a Nuclear Rocket

Currently thermal nuclear rockets seem to be the best choice for initial use in long space journeys in the solar system. But Pulsed Pellet Micro-Nuclear Explosion Rockets and Pulsed Pellet Micro-Nuclear-Fusion Explosion Rockets are worthwhile projects to pursue on a small scale as potential second and third generation nuclear rockets. Their R&D offers the additional bonus of being similar to aspects of the propose starship described in later chapters.

The obvious current priority choices for journeys by nuclear rocket are to Mars, the Asteroids, the moons of Jupiter and Saturn, and Venus. We place Venus last because of its size, high temperatures, and the low probability of finding life or interesting materials there. At best Venus will allow us to "test" theories of planetary dynamics on a planet of similar size to earth.

4.8 Benefits of Nuclear Rocket Program

In addition to the creation of a fast long range nuclear rocket that will be able to explore the Solar System in a cost effective manner there are a number of other benefits likely to result from the R&D effort to construct it. Among these benefits are:

1. Improvements in laser technology.
2. Improvements in nuclear power generation technology and possibly the development of new micro-nuclear powered devices that would replace batteries.
3. Contributions to fusion power technology.
4. Improvements in magnet technology.
5. Exploration of other planets and moves might improve our understanding of the earth's dynamics leading to better knowledge of earthquakes and volcanos.
6. The R&D effort would provide meaningful employment for the most educated technologists as well as the workers who build the rocket – a powerful motivation for young people to become educated.
7. It would provide a national goal in a time of uncertainty, "class warfare", and disillusionment.

4.9 Visualizations of a Possible Nuclear Rocket

There are many possible designs for nuclear rockets. They could look like conventional rockets with the thrust emanating from the tail of the rocket. This design of a nuclear rocket seems appropriate for thermal nuclear rockets.

In the case of Pulsed Pellet Micro-Nuclear Explosion rockets a circular design appears more appropriate. The circularity follows from the need for circular accelerators for the fuel pellets. The outer part of the vehicle could be the circular accelerator(s) (with magnets to direct the pellets) as depicted in Fig. 4.2. The inner parts would hold the crew, fuel and cargo as well as nuclear reactors for power. Fig. 4.2 shows one of the points where the pellets collide and generate thrust. It seems that 2, 4, or many such thrust generation points (Fig. 4.3) could encircle the craft and power the rocket. The choice depends on detailed design considerations of efficiency and maneuverability.

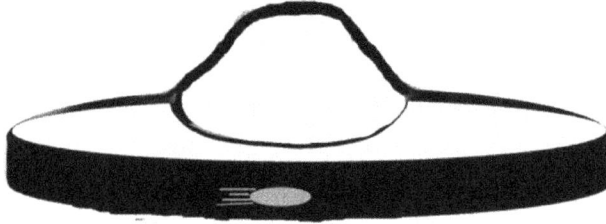

Figure 4.2 Schematic design of Pulsed Pellet Micro-Nuclear Explosion rocket with one of two diagonally opposite pellet thrusters shown.

Figure 4.3 Schematic design of Pulsed Pellet Micro-Nuclear Explosion rocket with many pellet thrusters shown. They are equally placed around the ring with variable thrust from each thruster for maneuverability. The thrusters combine to produce a large cumulative total thrust.

4.10 Recommended Solar System Rocket Development Program

The major nuclear rocket program for our solar system should be a nuclear thermal rocket along the lines of the NERVA program and its successors – especially because it offers the possibility of refueling propellant on outer planets' moons. In addition there should be an R&D program for the Pulsed Pellet Micro-Nuclear Explosion rocket to see if it is a more cost-effective alternative.

5. Long Shelf Life Nuclear Rockets for use in Extra-Solar Planetary Systems

Later in this book we will consider starships that travel faster than the speed of light with the consequence that time passes much more quickly than on earth. If the starship speed is hundreds or thousands or more times the speed of light then the passage of time on the starship is approximately the speed of the starship times the passage of time on earth.[25] Table 5.1 shows the relation between starship time and earth time for a starship traveling at 5000 times the speed of light.

Distance	20 Light Years	5000 Light Years	100,000 Light Years	2,000,000 Light Years
Earth Time	1.5 days	1 year	20 years	400 years
Starship Time	20 years	5000 years	100,000 years	2,000,000 years

Table 5.1 The relation between earth time periods and starship time periods for a starship traveling at 5000 times the speed of light. Elapsed earth travel time vs. starship travel time.

Table 5.2 shows the relation between starship time and earth time for a starship traveling at 1000 times the speed of light. Note that the expended starship time numerically equals the distance traveled if we use years and light years as units respectively. The expended earth time varies with the distance: it is the ratio of the distance to the speed.

---

[25] We will discuss this topic in much more detail in subsequent chapters.

Distance	20 Light Years	5000 Light Years	100,000 Light Years	2,000,000 Light Years
Earth Time	7.5 days	5 years	100 years	2000 years
Starship Time	20 years	5000 years	100,000 years	2,000,000 years

Table 5.2 The relation between earth time periods and starship time periods for a starship traveling at 1000 times the speed of light.

From Table 5.1 we see a 20 light year trip would take 1.5 earth days (not counting the time required to accelerate to 5000 times the speed of light) but it would appear to be 20 years on the starship. For extremely long journeys the starship travel time would be far larger. As a result trips of great length will necessitate suspended animation for the crew and extremely long lifetime equipment and machinery to survive the time period. That will enable the crew to maintain roughly equal physiological time to elapsed earth time.

One of the necessary requirements for a starship is a shuttle – presumably a nuclear shuttle – that could land on planets and moons of remote stars. Thus we need to consider how to build nuclear ships that can be inoperative for very long times and then start at the end of a journey to a star – "long shelf life" nuclear rockets.

5.1 Shelf-Life Requirements for the Nuclear Shuttles of a Far Traveling Starship

The nuclear rockets that we discussed in the previous chapter are sufficient for use as shuttles for starships that travel to stars in our "neighborhood" (approximately 20 light years – a region containing about 100 stars and a growing number of earth-like planets) but not usable for travel to distant stars and galaxies due to the accelerated time on fast starships. We can summarize the suitability of our Solar System nuclear rockets for various distances at speeds of 5,000 times the speed of light by:

- Distances Up To 20 Light Years
    Nuclear Rockets of the type of the preceding chapter

- Distances Up To 5,000 Light Years
  Long Shelf Life Nuclear Rockets

- Distances Up To 100,000 Light Years
  Long Shelf Life Nuclear Rockets

- Distances Up To 2,000,000 Light Years
  Long Shelf Life Nuclear Rockets
  Within this distance a starship can travel to numerous galaxies – some small satellite galaxies – but also larger ones such as the Magellanic Clouds, and the Andromeda Galaxy at a distance of 2,000,000 light years

5.2 Long Shelf Life Nuclear Reactors

Long shelf life nuclear rockets are nuclear rockets that can remain dormant for up to many thousands or millions of years before being used after the target star is reached. In the initial stages of starship exploration, within a 20 light year radius of our sun, nuclear shuttles of the type described in the preceding chapter will suffice – perhaps with some modifications.

However a long term project to develop long shelf life nuclear shuttles should be started after the starship R&D confirms the possibility of faster than light starships. The costs of this project, which is intimately tied to the development of long-lived metal alloys and ceramics, can be spread over a period of years to minimize the yearly expenditures.

It is difficult to provide a concrete design for a long shelf life nuclear rocket due to present uncertainties in the materials that would be used to construct the nuclear reactor(s) to power the shuttle and to generate thrust. More importantly the design of the reactor(s) itself remains to be developed. It must conserve fissile materials for long periods of time with only natural decay taking place. Then the fissile materials must be "brought" together to constitute a working reactor(s). There are a number of fissile materials (Uranium, Plutonium, and Thorium isotopes) with long half-lives that would be suitable depending on the travel length of journeys.

We assume a starship runs on minimal power during the coasting phase after it reaches its cruising speed. Near its destination the starship begins a boot up process that first slows the starship down to a few miles per second and starts up nuclear reactors on the starship, and on its shuttles, for use at the destination when the crew is revived from suspended animation (discussed later).

The startup process for a long lifetime nuclear shuttle or starship nuclear reactor is:

1. An "alarm" on a long lived clock on the starship goes off and a nuclear reactor bootstrap process begins on each shuttle of the starship, and for the nuclear reactors on the starship that generate power to run the starship and provide a backup nuclear engine(s) to propel the starship within the destination solar system.
2. The bootstrap process involves starting increasingly larger nuclear power sources from perhaps a low power thermocouple nuclear power source. The larger nuclear power sources may mechanically assemble the parts of the large reactors that are used by the starship and its shuttles. They decelerate the starship, and provide power for the starship ion engines and its shuttles nuclear engines at the destination. We envision a starship to have nuclear engines as well as quark-gluon ion drive engines.

The preceding discussions do not describe our starship designs. We describe these topics in subsequent chapters and mathematical appendices.

5.3 *Goals for Long Shelf Life Nuclear and Chemical Rockets*

Long shelf life nuclear and conventional rockets that could be dormant for 20 – 1000 years have a set of necessary requirements:

1. They must be able to start after dormant periods of up to 1,000 years.

2. They must be reliable in operation after being started.
3. They must be able to be shut down for periods up to 1,000 years and then restart.
4. They must have a long operating life under repeated restarts.
5. They must be efficient.
6. They must be able to land on up to earth size planets (and also moons) and lift off to return to the starship. They must produce minimal radiation during landing and takeoffs from planets and moons.

5.4 Approaches to Automatically Activating and deactivating Nuclear Reactors

For extraordinarily long starship time journeys we have seen the need to keep nuclear reactor fuel dormant except for the normal processes of decay. (Many fissile elements have half-lives greater than half a million years and so would be suitable fuel.)

Near the end of a trip the starship would need to assemble reactor components in such a way as to begin normal reactor operation and power generation. In addition after operation in a remote star system the nuclear reactors would need to disassemble to a dormant state again. They would have to be reassembled after travel to another solar system. The startup and shutdown process must be safe and completely reliable upon repeated reuse.

These requirements put limits on the types of reactors that one can automatically assemble and disassemble. There appear to be two types of nuclear reactors with a simple (thus more reliable) activate-deactivate capability:

1. Pebble Reactors
2. Compressible gaseous reactors

Each of these types has the fissile material separated while dormant so that only natural radioactive decays occur in the separated fissile material parts. When the fissile material parts are "brought together" than the resulting "assembled" reactor reaches the critical level for sustained nuclear reactions and energy generation. When the reactor must be deactivated for a dormant period the parts are then "separated" and revert to a state of natural decay.

5.5 Pebble Reactors

The concept of pebble reactors was first proposed in 1947 by Prof. Farrington Daniels at Oak Ridge. Since then a number of pebble reactors have been constructed in Germany, South Africa, and most prominently in China which is engaged in the construction of 30 pebble reactors by 2020 to produce 6 gigawatts of electric power.

The main value of pebble reactors for our purposes is the ability to turn the reactor "on" and "off" in a conceptually simple manner. Thus we can activate or deactivate a nuclear rocket in a safe, reliable way – although some technical hurdles remain to be overcome. The experiences with the European pebble reactors and the large number of Chinese pebble reactors in the design, testing and construction stages over the next 10 years should lead to the design of a reliable pebble nuclear rocket.

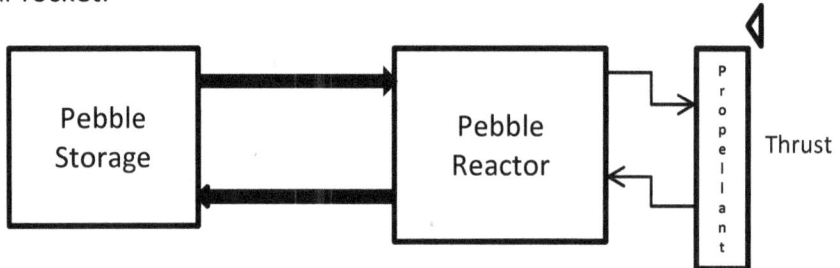

Figure 5.1 Restartable pebble bed reactor. When dormant, the pebbles are stored in a Pebble Storage container immersed in a neutron absorbing fluid that prevents criticality and chain reactions. When started, pebbles are loaded into the reactor (but not the storage container fluid). The reactor begins operating after reaching critical mass. The reactor contains helium or another non-neutron absorbing fluid. This fluid transfers heat to the propellant. The heated propellant generates power or thrust. The reactor is brought back to a dormant stage by transporting pellets back to the storage container. The storage container contains a neutron absorbing fluid that reduces fission within pellets to normal radioactivity. Since natural radioactivity generates heat the storage container can itself generate some electricity while the

rocket is "on the shelf." This electricity can power devices on the rocket (and starship) while in transit.

Fig. 5.1 illustrates the general design of a pebble nuclear rocket engine. Pebbles are roughly the size of tennis balls and currently are usually composed of pyrolytic graphite that each contains thousands of micro-particles of coated fissile material called TRISO particles. TRISO particles consist of fissile matter coated with silicon carbide ceramic to maintain the structural integrity of the pebbles.[26] The power generation potential of a chamber (of the right geometry) containing pebbles is indicated by the generation of 120 megawatts of power by a reactor chamber containing roughly 400,000 pebbles – each containing about 9 grams of U-235. The details of the fuel and structural materials remain to be determined. The propellant will initially probably be hydrogen. But the ship should also support refueling using other liquid fuels such as methane that naturally occur on many moons and planets.

Thus pebble reactors, which are quite safe and reliable, also have a simple activate/deactivate procedure that makes them suitable for shuttles on far traveling starships.[27]

5.6 Compressible Gaseous Reactors

Another type of simple long shelf time nuclear reactors is a gaseous reactor that is activated by compressing the container holding a fissile gas to a size where the gas density makes it become critical and begin to generate energy through fission. The gaseous reactor can be deactivated by expanding the container to the point where the fissile gas density does not support a critical nuclear chain reaction. The dilute gas particles then decay "slowly" through natural radioactivity. Tubes extending through the container heat the propellant and generate power or thrust. The propellant is flowed through tubes embedded in the reactor, is heated to high temperature, and exits to provide rocket

---

[26] The ultimate composition and shape of the pebbles remain to be determined from earth-based pebble reactor experience and R&D studies. Pebble reactors have a substantial history since 1947 that the reader may wish to investigate.

[27] Earlier pebble reactors were found to have problems such as deformed pebbles, clogging of pellet movement and so on. None of these problems is insurmountable as witnessed by the decision of the Chinese government to build a large number of pebble reactors.

thrust or power. Typically the propellant is hydrogen. But the design could support methane and other propellants so that propellant refueling becomes possible on distant moons and planets.

The simplest geometry of the reactor container would be a cylinder with a piston at one end that could compress or decompress the fissile gas within the container. Tubes in the container heat the propellant from the fuel tank as it flows through the container. See Fig. 5.2 for a schematic view of the container.

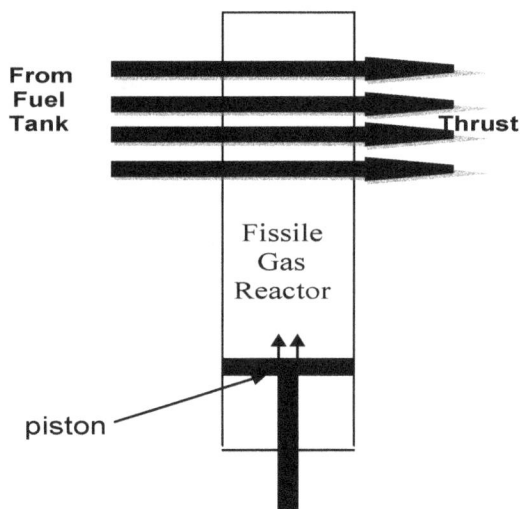

Figure 5.2 Schematic depiction of the cross section of a cylindrical fissile gas container with a piston. The total thrust is the combination of the thrust generated from each pipe going through the reactor.

Fissile gas reactors have been a subject of investigation for some time. They are often called Gaseous Fission Reactors. The main problem of this type of reactor is the corrosive effects of the fissile gas which is typically UF_4 diluted in helium in current terrestrial reactors. Other fissile gases are possible such as uranium hexafluoride. A key determining factor for the choice of fissile gas is its corrosiveness. A less corrosive fissile gas will give a longer reactor lifetime.

The reactor can be simply started and stopped thus providing a long "shelf life" reactor. It is started by compressing the dilute fissile gas to a critical density when a chain reaction can start, and propellant can be pumped through the reactor to generate thrust or power. After running for the required time in a remote solar system the reactor can be brought to a dormant stage by withdrawing the piston to the point where the fissile gas is too dilute to engage in a chain reaction. The fissile gas then reverts to normal radioactivity. Since natural radioactivity also generates heat the dormant reactor can generate some electricity while the rocket is "on the shelf." This electricity can power devices on the rocket (and starship) while in transit to the destination solar system.

5.7 Conclusion

These first five chapters describe slower than light travel and transport using new types of space guns and rockets. They would enable a much more cost effective and efficient approach to space endeavors. In the next chapter we show there is good reason to believe that faster than light transportation is possible. The following chapters describe features of starships and faster than light travel. In the appendices we show how faster than light starship propulsion is feasible using a quark-gluon ion drive. The discussion in the appendices is necessarily mathematical. The non-mathematical reader can skim them to get the flavor of starship dynamics.

6. Experimental Evidence for Faster-Than-Light Particles & Physics

Until 1907 physicists thought that there was no limit on the speed of a particle or lump of matter. In 1907 Einstein and Poincaré showed that there was an inherent limit on the speed of a massive object – the speed of light. For the past 100 odd years physicists have generally accepted the speed of light as the limiting speed for particles with mass. Several theoretical physicists in the 1960's (E. C. Sudarshan and Gerald Feinberg) investigated the possibility of faster than light particles. They found that faster than light particles were theoretically possible but their theories – particularly their quantum field theories – had numerous discrepancies from canonical quantum field theory. These differences were taken by many to indicate that faster than light particles (called tachyons) were not present in nature. This belief was further supported by the happenings at particle accelerators where it was impossible to accelerate normal charged particles faster than the speed of light.

In the past ten years this author[28] developed a satisfactory theory of faster than light particles and found that if neutrinos and quarks were faster than light particles he could derive the form of The Standard Model of Elementary Particles in detail. This theoretical development seems to have stimulated experimental groups at the new Linear Hadron Collider (LHC) at the CERN laboratory in Switzerland and the Gran Sasso Laboratory in Italy to measure the speed of neutrinos emitted in LHC particle collisions. The results, described below, were mixed and one can fairly say they neither proved nor disproved that neutrinos were tachyons.

---

[28] See Blaha (2012b) and earlier books extending back nine years.

However there is other experimental data that strongly indicate that neutrinos are tachyons and that quantum mechanics requires – not just faster than light behavior – but in some circumstances instantaneous effects at a distance – infinite speed of transmission!

In this chapter we will look at experimentally proven instantaneous Quantum Mechanical effects, tritium decay experiments over the past 20 years implying faster than light neutrinos, neutrino speed measurements at the LHC and Gran Sasso, tachyonic particle behavior inside of Black Holes, and the tachyonic behavior of Higgs particles, the "so-called God particle," (should they exist in nature as recent experiments suggest.) *The cumulative result of these considerations is that faster than light particles, and physics, are part of nature.*

6.1 Instantaneous Quantum Mechanical Effects

Quantum entanglement is a quantum phenomenon wherein parts of a physical system are in a certain quantum state but are separated by a space-like distance. If a change is made in part of a quantum entangled system then it is known theoretically, and experimentally, that other parts of the system change instantaneously.[29] Many experiments have shown that the change in other parts of a system is instantaneous and thus can be viewed as taking place at infinite speed – obviously beyond the speed of light.[30] These experimental results are consistent with quantum mechanics. Thus faster than light behavior is implicit in quantum theory.

6.2 Tritium Decay Experiments Yielding Neutrinos

We note that particles with negative values for the square of their mass are tachyons – particles moving faster than light.

A series of experiments by various groups over recent years imply that electron neutrinos produced in tritium decay have negative

---

[29] Matson, John, "Quantum Teleportation Achieved Over Record Distances" *Nature* **13**, August 2012.

[30] Francis, Matthew, "Quantum Entanglement Shows that Reality Can't be Local", *Ars Technica*, 30 October 2012.

mass squared despite the best efforts of experimenters to obtain positive values for the mass squared.

Experiment	measured mass squared	Year
Mainz	-1.6 ± 2.5 ± 2.1	2000
Troitsk	-1.0 ± 3.0 ± 2.1	2000
Zürich	-24 ± 48 ± 61	1992
Tokyo INS	- 65 ± 85 ± 65	1991
Los Alamos	- 147 ± 68 ± 41	1991
Livermore	- 130 ± 20 ± 15	1995
China	- 31 ± 75 ± 48	1995
1998 Average	-27 ± 20	1998

Table 6.1 Electron neutrino mass squared values found in various tritium decay experiments. (Masses are in units of eV.) The average mass squared is negative suggesting electron neutrinos are tachyons.

Table 6.1 summarizes the measured electron mass squared in these experiments. These experiments strongly suggest that neutrinos have negative mass squared and are thus faster-than-light particles - tachyons.

6.3 LHC/Gran Sasso Direct Measurements of Neutrino Speeds

Two groups performed experiments at Gran Sasso Laboratory in Italy. They detected neutrinos emitted in interactions at the CERN LHC in Switzerland. The LVD collaboration in an exhaustive study of neutrino velocities found that the question was open according to their data. Their refereed Physical Review Letter Abstract stated:

We report the measurement of the time of flight of ~17 GeV v_μ on the CNGS baseline (732 km) with the Large Volume Detector (LVD) at the Gran Sasso Laboratory. The CERN-SPS accelerator has been operated from May 10th to

May 24th 2012, with a tightly bunched-beam structure to allow the velocity of neutrinos to be accurately measured on an event-by-event basis. LVD has detected 48 neutrino events, associated with the beam, with a high absolute time accuracy. These events allow us to establish the following limit on the difference between the neutrino speed and the light velocity: $-3.8 \times 10^{-6} < (v_\nu - c)/c < 3.1 \times 10^{-6}$ (at 99% C.L.). This value is an order of magnitude lower than previous direct measurements.[31]

These results (involving at least 35 neutrino detections) slightly favor, and do not rule out, faster-than-light neutrinos. Another experiment at the same locations by the ATLAS group stated that they found neutrino velocities (5 neutrinos were measured) were below c. This group has not published their results as yet. We conclude that the published data appears to support faster than light neutrinos – consistent with our theory of The Standard Model.

A new project is in the planning stages to measure neutrino beams at larger distances. The hope is that the masses of the various neutrinos will be determined by the experiment. If the neutrino mass squared values turn out to be negative then it will constitute additional proof that neutrinos are tachyons (confirming tritium decay data), and thus support this author's formulation of The Standard Model of Elementary Particles.

6.4 Tachyonic Behavior Within Black Holes

Inside a black hole (such as the Schwarzschild solution of General Relativity) the time coordinate effectively becomes a spatial coordinate and the radius coordinate effectively becomes a time coordinate. An in-falling particle has a constantly decreasing radial distance from the center of the black hole just as time always increases outside a black hole.

As a result of the interchange of the roles of time and radius the velocity of a particle descending radially inside a Black Hole has a speed faster than light and is tachyonic.

---

[31] N. Yu. Agafonova et al. (LVD Collaboration), "Measurement of the Velocity of Neutrinos from the CNGS Beam with the Large Volume Detector" Phys. Rev. Lett. **109**, 070801 (15 August 2012).

6.5 Higgs Fields are Tachyons

Recently groups at the LHC CERN laboratory have announced the discovery of Higgs particles. The dynamic equations for Higgs bosons in The Standard Model have a negative mass squared. The mass squared must be negative or the Higgs Mechanism could not generate particle masses. Having negative mass terms implies that Higgs fields are tachyonic – faster than light particles. Their tachyonic nature is masked by a quartic self-interaction that generates a condensate and thereby the masses of other particles.

6.6 Conclusion: Faster-Than-Light Particles – Tachyons Exist in Nature

The bulk of the experimental and theoretical evidence presented in previous sections favors the existence of faster-than-light particles such as neutrinos. Tachyonic neutrinos are an important part of our form of The Standard Model. This form of the theory also strongly suggests that quarks are tachyonic in parallel with tachyonic neutrinos in order to obtain the symmetries of The Standard Model.

6.7 Faster than Light Starships are Possible Using Tachyon Thrust Drives

Thus we can sensibly consider the possibility of using tachyonic quarks for an ion starship drive. We shall see that a quark-gluon plasma can provide a faster than light thrust to drive a starship to speeds in excess of the speed of light. In particular, we shall see that a starship, so driven, can evade ("go around") the barrier at the speed of light.

The mathematics of a faster than light starship is described in appendix B. The non-mathematical reader is not required to read this appendix to understand the general concepts behind a faster than light starship. The following chapters describe features and requirements for faster than light starship travel in non-mathematical terms.

7. The Only Hope for Humanity: Cheap, Faster-Than-Light Migration to the Stars

As the world population grows and the overall environment deteriorates it is important to look to the possibility of large scale migration to other earth-like planets over the next millennium. In this chapter we consider some aspects of migration.

7.1 Many Other Solar Systems – Some with Earth-like Planets

At the time of this writing new observational tools and techniques are starting to reveal the presence of earth-like (habitable) planets within a relatively short distance of our sun. The requirements for earth-like planets include an orbit around a suitable star in a zone where liquid water can exist on a planet, a rocky planet (as opposed to a gas giant planet like Jupiter), and a planet with a mass comparable to that of earth up to perhaps 8 times earth's mass.

As of December 5, 2012 the Habitable Planets Catalog lists 7 habitable planets and 27 candidates for habitable planet status. The seven planets in the Catalog are Gliese 581 g (mass 2.6 earth masses, distance 20.2 light years, 1.4 times earth's radius), Gliese 667C c (mass 4.9 earth masses, distance 23.6 light years, 1.9 times earth's radius), Kepler 22 b (mass 2.6 earth masses, distance 20.2 light years, 1.4 times earth's radius), HD 40307 g (mass 6.4 earth masses, distance 539.9 light years, 2.1 times earth's radius), HD 85512 b (mass 4 earth masses, distance 36.3 light years, 1.7 times earth's radius), Gliese 163 c (mass 8.3 earth masses, distance 48.9 light years, 2.4 times earth's radius), and Gliese 581 d (mass 6.9 earth masses, distance 20.2 light years, 2.2 times earth's radius).

In addition Alpha Centauri, only about 4 light years away is now known to have at least one earth-like planet. The three Centauri stars are likely to have more earth-like planets – some of which may be in the habitable regions around these stars. Earth-like planets are also very likely to exist circling other stars in the "neighborhood" of earth. Tau Ceti which is 11.9 light years away appears to have at least two earthlike planets circling it in its habitable zone – one planet is four times the size of earth and has a 6 month year.

The galaxy is estimated to have 17 billion planets of roughly earth-like size based on results from NASA's Kepler Space Observatory.

7.2 *Percentage of Habitable Planets That are Truly Earth-like*

It is not possible to estimate the number of planets of earth-like size in a solar system in its habitable zone where liquid water is possible, and having an atmosphere and water content suitable for life as we know it. However if we take the position that Nature is fairly consistent in its phenomena, then, if our solar system is a reasonable guide, we can expect perhaps one in three earth-like size planets in habitable zones to have an environment conducive to life. Our Solar System has three potential earth-like planets: Mars, Earth and Venus. Only Earth is "well equipped" to support anything more than microbial life.

Thus of the 17 billion earth-like planets it seems reasonable to conservatively anticipate that there will be of the order of 16 billion earths. If that estimate is anywhere near true then we can expect to find many planets capable of supporting human-like life. We can also expect that many planets developed other forms of life including intelligent life. Venturing into the Cosmos has both benefits and potential dangers.

A starship should thus have strong armaments should the need arise ranging from side arms to nuclear guided missiles. We come in peace but must be prepared for other eventualities. A possible major problem is communication with other species. We can establish only the most rudimentary communication with "semi-intelligent" species on earth. Communicating with intelligent aliens in an alien environment would be significantly more difficult.

7.3 Feasibility of Other Starship Proposals to Travel to the Stars

There are two modes of travel to far star systems: faster than light starships and multi-generation slower than light starships.

7.3.1 Faster Than Light Starships

There are two proposed approaches to faster than light starships – the quark-gluon ion drive starships originally proposed by this author and starships based on gravitational effects: black holes, worm holes, gravitational waves such as Alcubiere drives, and so on. Gravitationally-based drives cannot succeed for two reasons: the masses required to achieve such effects are so large as to be beyond the resources of humanity (well over Jupiter's mass); and a crew in such a drive environment, if one could be created, would be crushed.

Consequently the quark-gluon ion drive is the only possible faster than light drive that is currently capable of being implemented with humanity's resources. There are major R&D hurdles to building a starship using this drive. But there is no apparent alternative possibility.

Gravitational drive studies may be of theoretical interest but they are impossible as a mechanism for starship drives within the framework of present or future physics, and the present or future capabilities of humanity.

7.3.2 Multi-Generation Starships

There are many proposals for starships that would require many generations to reach even nearby stars: solar sails, ships powered by nuclear or hydrogen bombs, and so on.

These proposals require ships of immense size, pose profound psychological issues for generations of crews, and are not fast enough to make human migration and trade between worlds – commerce – possible. The result at best would be to plant small human colonies in isolation in remote environments that are not known.

Thus starships of this sort are not of practical interest for the purpose of establishing a widespread, interacting multi-star human community.

We conclude quark-gluon ion drive starships offer the only means of creating a dynamic, galactic human civilization.

7.4 Short Trips to Nearby Stars

It is comforting to know that there are nearby stars that have earth-like planets. While there is no guarantee that any have a truly earth-like planet with water, an earth-like atmosphere and a "comfortable" range of temperatures; there are earth sized planets in the habitable region around their stars. And this recent finding of the ongoing search for planets gives us good reason to start the effort to travel to nearby stars.

7.5 Nearby Stars with Earth-like Planets

At the time of this writing (January, 2013) the most promising nearby possibilities for earth-like planets in the habitable zone of sun-like stars are planets of the three Centauri stars at approximately 4 light years distance and of Tau Ceti at a distance of approximately 10 light years.

Since there are approximately one hundred stars within a 20 light year radius of earth there are undoubtedly other earth-like planets within that distance.

7.6 A Trip to the Centauri Stars or Tau Ceti

We can estimate the general characteristics of a trip to these nearby stars. Starting from the vicinity of earth the starship engines will accelerate the starship to a speed of perhaps 40 times the speed of light. The acceleration time will be several months. (We consider a detailed example of acceleration in the appendices. Since we exceed the speed of light relativistic effects strongly affect the dynamics of acceleration.) After reaching the target speed the ship "coasts" to near the Centauri stars. Then the starship decelerates to a speed of a few miles per second. The starship's nuclear engines then take over and the extended solar systems of these stars can be explored.

Nuclear shuttles can make close approaches to planets and moons of interest. They may choose to land on some of these bodies to search for life and explore their geology.

Depending on the starship and the availability of propellant on bodies in these solar systems, the starship may take advantage of the opportunity to refuel, although the starship should have sufficient propellant to return to earth without refueling. Refueling would give the starship the ability for extended exploration.

After the exploration phase is finished the starship will accelerate to cruising speed for the return trip to earth. After the cruising phase the starship will decelerate to a speed of a few miles per second and use its nuclear engines to return to earth orbit (or perhaps an orbit around the moon if a moon colony exists that can analyze the starship's findings.)

8. Breaching the Light Barrier - Quark-Gluon Plasma Ion Drives

8.1 Two Forms of Quark-Gluon Plasma Ion Drive

There are two simple forms for a quark-gluon plasma ion drive. This drive consists essentially of a chamber that contains a plasma of great temperature and pressure consisting of quarks (the particles of which elementary particles such as protons and neutrons and so on are composed), and gluons (particles that carry the strong force – quarks are usually bound together in protons, neutrons, and so on by this force).

Quark-gluon plasma is created by colliding atoms or spherical clumps of atoms which we call *spherules* at extremely high energy in accelerators. The leading accelerator today for creating quark-gluon plasma is the Linear Hadron Collider (LHC) located at the CERN laboratory in Geneva, Switzerland.

The quark-gluon plasma created by a collision at the LHC immediately expands explosively to become a stream of "normal" particles through the interactions and combinations of the quarks and gluons. If we could confine the plasma in an enormously strong magnetic "bottle" so it could not expand then the plasma could not break up into "normal" particles. Then if we open the bottle a bit to allow the quark-gluon plasma to squirt out we would have a quark-gluon plasma ion drive.[32] See Fig. 8.1 for an illustration of the bottled thrust process.

What advantage does a quark-gluon ion drive have? The thrust of each particle in the stream of quarks and gluons has complex[33]

---

[32] Normal rocket ion drives use charged particles and atoms as the ions in the rocket thrust.

[33] Complex means that the quantity is in part a real number and in part an imaginary number (a number that is a multiple of the square root of minus one.)

spatial momentum. This stream imparts a complex acceleration to the containing bottle (which of course would be part of a starship) and thereby to the starship that it propels. A starship with a complex acceleration, which generates a complex velocity, can "drive" around the speed of light – a real number – and attain faster than light motion.

There is a secondary effect here that deserves mentioning. After the particles exit from the bottle with complex momenta they can combine to make ordinary particles with strictly real momenta and also "suck" particles out of the vacuum, combine with them, and make normal particles with real momenta. This secondary process that takes place after exiting the bottle (and starship) creates a drag such as is seen in aerodynamics. But the drag does not slow the starship because it happens after the exit from the starship.

The net result is that we have a quark-gluon plasma ion drive starship that can exceed the speed of light. And so we can go quickly to the stars![34]

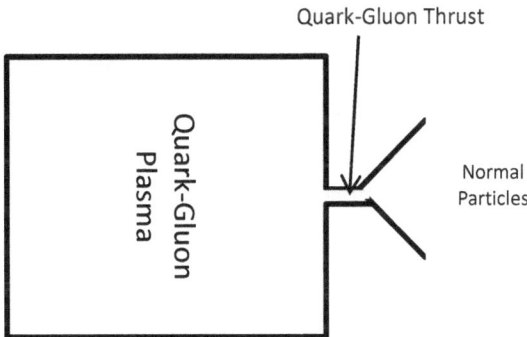

Figure 8.1 Magnetic bottle containing quark-gluon plasma. Plasma creation by high energy collision of atoms or spherules is not shown. The plasma exits, creating thrust which afterwards converts through interactions to normal particles.

---

[34] This scenario presupposes that we can build magnetic bottles that can contain "large" quantities of quark-gluon plasma. There are significant R&D challenges to be met to achieve that goal. But it does not appear to be precluded by physical principles if quarks and gluons have complex momenta – a key factor in our derivation of The Standard Model of Elementary Particles. See Blaha (2012b) and earlier work referenced therein.

8.2 Accelerators to Make Quark-Gluon Plasma

There are two general types of accelerators, circular accelerators and linear accelerators, that can accelerate atoms such as U-238 or spherules containing milligrams or micrograms of atoms to very high energies of the order of 200 GeV per nucleon.[35]

The two types of accelerators that could raise atom and spherule energies to the point where collisions could produce quark-gluon plasmas are:

1. Circular accelerators that accelerate atoms and spherules to high energy by repeated rotation in a circle.
2. Linear accelerators that accelerate atoms and spherules in a straight line.

Magnets can be used to divert the accelerated particles or spherules to collide with such force that the colliding objects are transformed into plasmas consisting of quarks and gluons. Quark-gluon plasmas have been created at the Brookhaven National Laboratory accelerator and more recently at the LHC. In these experiments the quark-gluon plasmas were not confined by a magnetic bottle and promptly interacted to become showers of normal particles. Presently sufficiently strong magnetic bottles to contain plasma are not available. But advances in superconducting magnet technology will hopefully make magnetic confinement of plasmas possible.

8.3 Circular Accelerator Engine

The general form of a circular accelerator engine consists of

1. a main accelerator ring(s), magnets to guide atoms or spherules around this ring, and
2. lower energy feeder rings to boost atoms or spherules to sufficient energy to insert in the main ring(s), and
3. intersection points where particles (atoms or spherules) collide to generate quark-gluon plasma. One cannot

---

[35] Nucleons are protons and neutrons – the particles that appear in the nucleus of atoms.

accelerate an atom or spherule effectively from zero speed to high energy in one ring. (Earth bound circular accelerators such as the LHC use several support accelerators[36] that consecutively boost particle energies to energies satisfactory for injection into the main LHC ring.)

4. magnets and a magnetic bottle to confine the quark-gluon plasma, and

5. an opening in the bottle with magnets to direct the quark-gluon thrust backward.

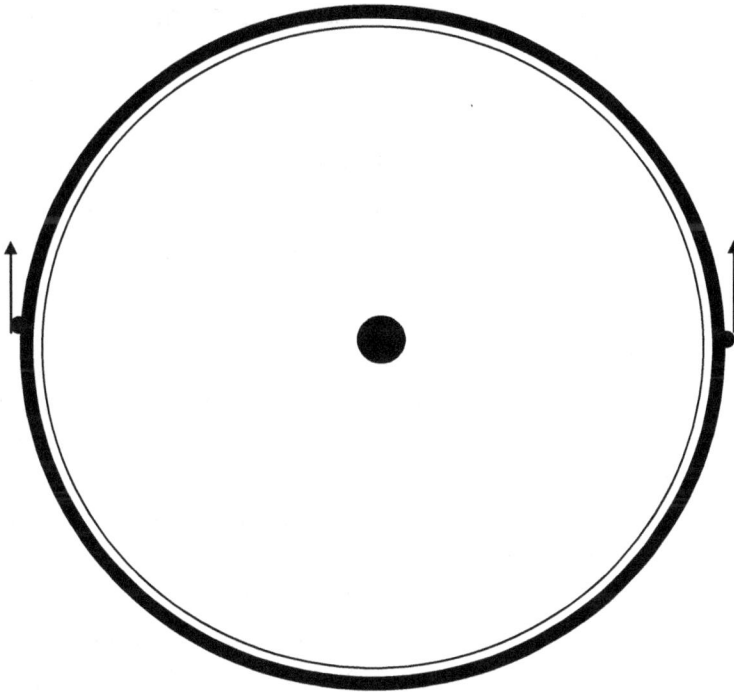

Figure 8.2 Top schematic view of basic circular accelerator engine with intersection points where interactions occur in a magnetic bottle with thrust indicated by small arrows. Fig. 8.1 is the form of the magnetic bottle indicated by a circle on each side of the

---

[36] Often the initial acceleration is provided by a linear accelerator.

accelerator ring(s). This configuration is suited to a saucer shaped starship with the starship "knifing horizontally" into the cosmos. If the bottles directed the thrust "downward" then the saucer would be moving face first into the cosmos. See Fig. 8.3.

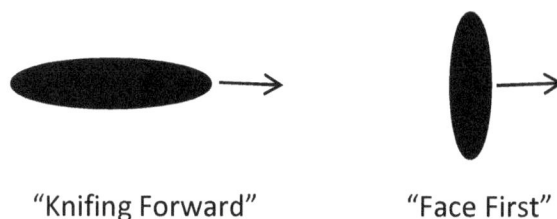

"Knifing Forward"   "Face First"

Figure 8.3 Types of motion for circular accelerator engined saucer shaped starships.

8.4 Linear Accelerator Engine

A linear accelerator engine consists of at least two linear accelerators that accelerate two (or more) streams of atoms or spherules to high energy whereupon the streams collide to produce quark-gluon plasma.

A linear accelerator engine consists of

1. Two or more linear accelerators[37] driven by electromagnetic fields in a linear fashion to boost atoms or spherules to high energy, and
2. lower energy feeder accelerators to boost atoms or spherules to sufficient energy to feed the linear accelerators, and
3. an intersection point where particle (atoms or spherules) streams collide to generate a quark-gluon plasma, and
4. magnets and a magnetic bottle to confine the quark-gluon plasma, and an opening in the bottle with magnets directing the quark-gluon thrust backward.

---

[37] A linear accelerator has a straight vacuum tunnel through which particles are accelerated and guided by electromagnets.

The linearity of the primary accelerator reduces the power required to drive this engine. However, unless new very rapid acceleration methods are developed (and there are some promising advances in this area) the

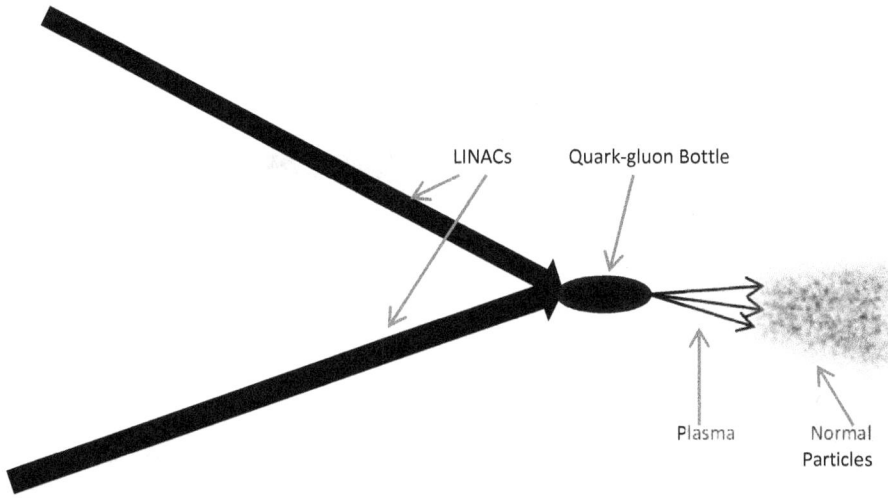

Figure 8.4 Schematic of a linear accelerator engine with two linear accelerators at an angle chosen to optimize the resultant quark-gluon thrust.

linear accelerators generally must be extremely long – of the order of many miles in length.

The general form of a linear accelerator engine (with two acceleration streams for the sake of illustration) appears in Fig. 8.4.

Engines with four linear accelerator streams are similar in configuration but the linear accelerators would be symmetrically arranged with two in a "North-South" arrangement and two in an "East-West" arrangement. All four streams would come to a common entry point of a magnetic bottle that confines the plasma. See Fig. 8.5.

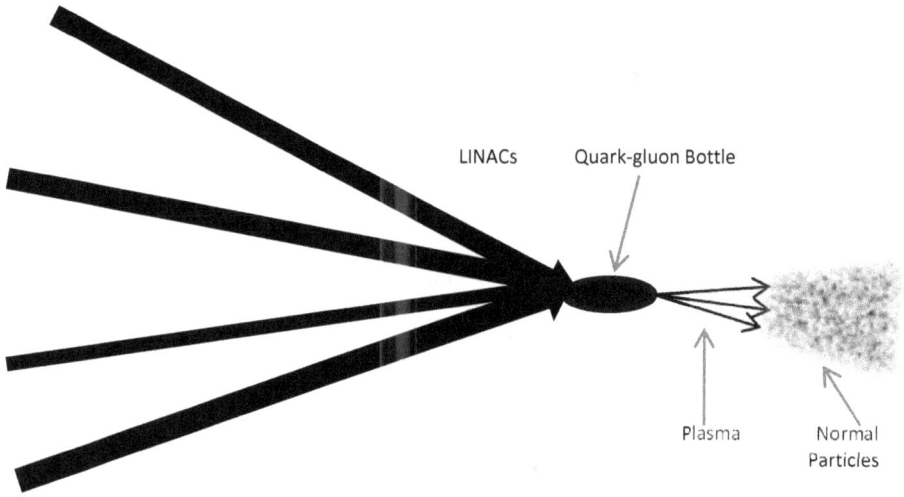

Figure 8.5 Schematic of a linear accelerator engine with four linear accelerators at opening angles chosen to optimize the resultant quark-gluon thrust.

9. Some Overall Designs of Starships

While one cannot predict the exact details of the shape of a starship, the mechanisms for creating quark-gluon plasma thrust suggest several possible configurations as reasonable choices. A consideration that we have not discussed as yet may also be of importance in starship design. *IF* space dust – the very low density of atoms and particles of matter between the stars – is a significant factor in the erosion and possible penetration of a starship's surface then consideration must be given to streamlining the starship's shape as well as protectively coating it with armor and possibly electromagnetic repulsing fields. The potential problem of space dust is magnified by the high faster-than-light speed of a starship. Particles of space dust under these circumstances would have a high penetrating power as well as a significant momentum that could cumulatively affect a starship's speed.

Figure 9.1 Circular Starship design with one starship thrust point on the edge of the ship. The engines below the ship disc are supplemental nuclear engines used for maneuvering.

Putting these considerations aside we will indicate several likely forms for a starship in this chapter. The effect of space dust can only be truly determined by prototype tests after an operational starship is constructed and tested. Detailed theoretical studies will also provide some guidance.

9.1 Starships Based on Circular Accelerators

Starships can be expected to have a circular overall form simply because the wider the main circular accelerator, the lower the energy requirements of the main accelerator bending magnets. See Fig. 9.1 for an example. The starship thrust is generated by pairwise combinations of complex thrust.[38] Fig. 9.2 illustrates the combination of the thrust of two quark-gluon globules to generate a complex rearward thrust.

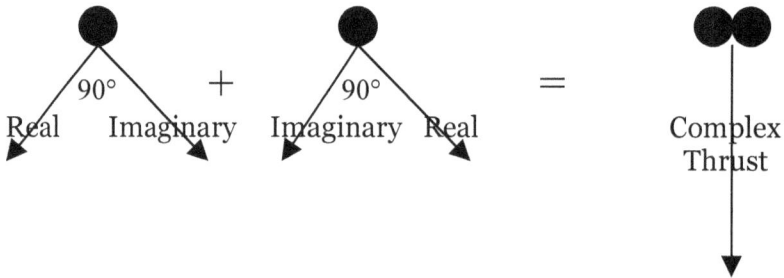

Figure 9.2 The combination of two quark-gluon fireballs to produce a complex thrust. This thrust can exceed the speed of light since the sum of the fireballs' momenta is a complex vector whose parts can exceed the speed of light.

---

[38] The complex thrust is a result of the sum of the complex velocity of the particles in the thrust. The velocity of the quark and gluon particles consists of perpendicular real and imaginary parts as shown in Fig. 9.2.

9.2 Cigar-Shaped Starships

A variation on circular accelerator starships is a cigar shaped starship that uses the imaginary part of the thrust to spin the cigar. This approach has the advantage of creating artificial gravity. However the imaginary part of the thrust must be varied repeatedly to provide a constant artificial gravity ("centrifugal force" – centripetal acceleration).

Figure 9.3 Cigar shaped starship with "horizontal accelerator ring(s). The real part of the thrust points downward. The imaginary part of the thrust is horizontal and tangent to the cigar surface. This part of the thrust causes the cigar ship to spin and generate artificial gravity (through centrifugal force – also called centripetal force). The imaginary part of the thrust must oscillate (using oscillating magnetic fields) to keep the gravity constant. The fins are for supplementary nuclear maneuvering rockets.

9.2.1 Artificial Gravity Levels - Rotating Circular and Cylindrical Starships

The real part of the thrust propels the starship. The imaginary part of the thrust causes the starship to spin around its central axis. The imaginary part of the thrust must oscillate quickly between clockwise and counter-clockwise to maintain a constant rotation speed (and thus constant values of gravity.) See the appendix H for a detailed discussion of the generation of artificial gravity by rotating discs and cylinders.

Top View of Disc or Cylindrical Starship
View of cargo/people levels of inner hub. The center has low "gravity." The outer parts have higher gravity. The "gravity" force is inward towards the center on all levels.

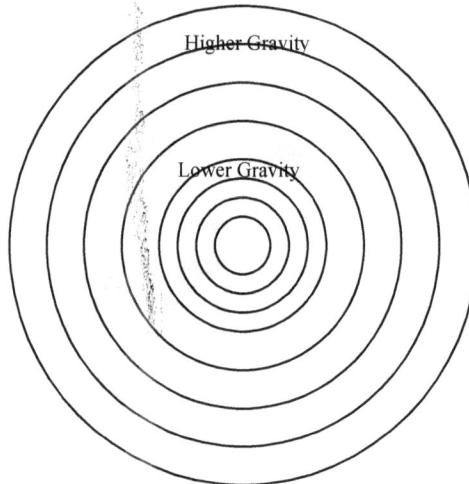

Figure 9.4 Levels of equal artificial gravity in a disc-shaped or cylindrical starship.

9.2.2 Rotating Circular Starship

A circular (disc-shaped) starship can also rotate to create artificial gravity. Fig. 9.5 illustrates a rotating disc starship, Again the gravitational pull is towards the center with stronger gravity in the outer region. The disc moves to the left due to the real thrust.

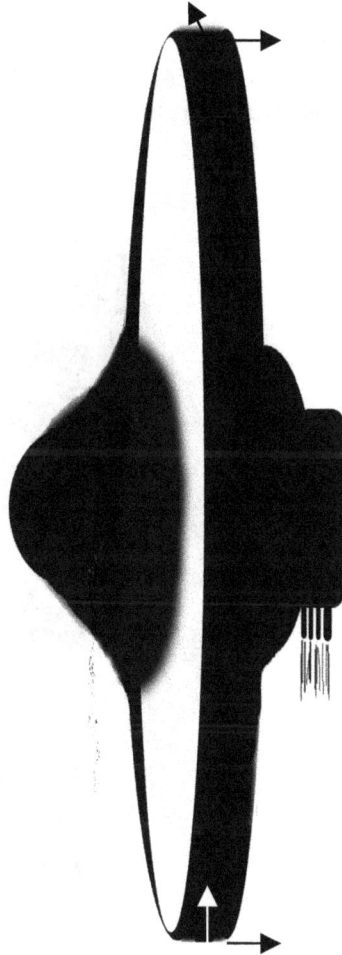

Figure 9.5 Rotating disc-shaped starship. The imaginary thrust causes rotation. The real part of the thrust is perpendicular to the disc. The imaginary part of thrust is tangent to the disc.

9.3 Starships Based on Linear Particle Accelerators

Starships based on linear accelerators can have a variety of forms. Fig. 9.6 contains a diamond shaped starship. Generally starships based on linear accelerators need great length – of the order of miles for the primary linear accelerator tubes unless a very rapid linear accelerator mechanism is developed. The angles between the linear accelerator tubes are determined by maximizing the efficiency, and amount, of thrust generation. Thus the shape of the lower half of the Fig. 9.6 starship is more or less determined. The ship can be sharp edged diamond shaped or rounded diamond shaped.

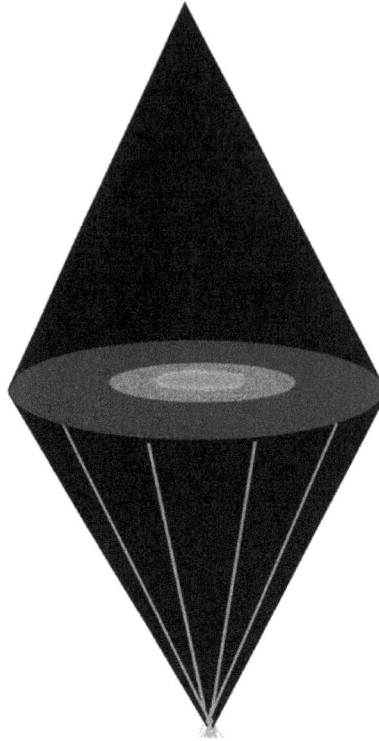

Figure 9.6 A diamond shaped ship powered by four linear accelerators. The lower part has the accelerators, magnets, nuclear reactors, and propellant. The upper part contains the crew, cargo, and shielding as well as nuclear shuttles for exploring a solar system.

10. Short Distance Starship Requirements

We will define a short distance starship as one that can travel to a star within a distance of about ten light years from earth. This distance, as pointed out earlier, includes about 100 stars, and the Centauri and Tau Ceti stars (which are known to have earth like planets in their habitable zones.) The requirements for these "short haul" starships are different from long distance starships. Long distance starships require a more advanced technology based on long lifetime materials, and suspended animation for crew members.

Short distance starships can be built with an incremental advance in technology (assuming quark-gluon ion drives can be built.) They do not need suspended animation for the crew.

10.1 R&D Stages

The construction of a short distance starship can proceed in stages in such a way that a failure at a critical stage can cause program termination with a minimal prior expenditure. Thus we can construct a starship in stages in the most cost effective manner whether the effort succeeds or fails at some stage. It must be realized that the construction of a starship requires technological advances that may or may not be realized. So taking account of risk factors at each stage is critical to proceeding in the most cost effective, least wasteful manner.

As we see it the stages of starship development are:

1. Detailed theoretical feasibility studies to determine whether a quark-gluon plasma ion drive can be built. This is a relatively low cost effort amounting to five million dollars spent over perhaps a five year period. It would be performed by physicists, engineers and mathematicians.

2. Assuming stage 1 is successive then a prototype(s) should be built on earth. This effort will perhaps take ten to fifteen years and cost perhaps twenty to thirty billion dollars spread over those years, or roughly three billion dollars per year.

3. The prototype should then be exhaustively tested. This effort should take two or three years and cost about three billion dollars or one billion dollars per year. Changes in the prototype resulting from test experiences should be expected.

4. If a prototype proves successful (another jeopardy point where project closure would be possible) then the design and development of a complete starship prototype should begin. The construction of the prototype should be in Far Earth orbit – perhaps 20,000 miles above the earth's surface to minimize any dangers of the prototype starship testing stage. The cost of this effort probably will be of the order of three hundred billion dollars spread over an estimated twenty year period or about fifteen billion dollars per year.

5. After construction test flights of the prototype in the Solar System to the outer planets should be performed for two years testing the maximum limits of the prototype. Cost estimate: four billion dollars for flying the constructed and fueled starship.

6. If successful the prototype should be brought to Far Earth orbit and expanded to a full sized starship for a trip to the Centauri stars. (It is assumed the prototype starship engines built in stage 4 would be full scale engines for the full sized starship. Cost estimate twenty billion dollars per year. Construction time: ten years.

7. A trip to the Centauri stars should follow.

A summary of the estimated length and cost[39] of each stage is:

Stage	Cost/year (billions $$)	Years
1	0.001	5
2	3	10
3	1	3
4	15	20
5	2	2
6	20	10
7	10	8

The total estimated cost is 617 billion dollars, and the total estimated construction and travel time of the full starship is 58 years. To those who fear the cost will hurt the American economy we say it will provide many jobs – but more than that it will provide a noble goal for the American people in a time of uncertainty and hopelessness.

10.2 Short Distance Starship Quark-Gluon Engines

Chapters 8 and 9 describe the general character of starship engines including both short distance and long distance starships. The primary difference between them is the amount of propellant and reactor fuel that they carry. The starship engine designs are similar.

10.3 Other Requirements

There are significant differences between short distance and long distance starships in other areas described in the following subsections.

---

[39] Current dollar estimates. These estimates are reasonable values meant to convey the cost and length of each stage. They are not necessarily accurate. Detailed cost studies should be made at the beginning of each stage of its cost and its execution time length. As in most R&D efforts costs often exceed estimates. For that reason our estimates were set to higher values than we expect to be the actual costs.

10.3.1 Nuclear Rockets

The first starship prototype could be built with conventional nuclear engines to lower the cost. Alternately the starship could be built with long shelf life nuclear engines for the purpose of testing this new type of engine in space. The starship shuttle(s), if built also, could use conventional nuclear engines since they would probably have a twenty year lifetime.[40] The decisions for the nuclear engines should be made on the basis of cost effectiveness at the time of construction.

Chapters 4 and 5 describe proposed new types of nuclear reactors and engines.

10.3.2 Possibly Short Term Suspended Animation

While a round trip to the Centauri stars can be accomplished in less than twelve years, it would be desirable if biomedical research into suspended animation would be strongly supported. If suspended animation for ten or more years becomes possible then it could be used on both short distance and long distance journeys. Suspended animation R&D could proceed in parallel with the stages of starship R&D. Some work in this area has already begun and, with stronger support, one can hope for significant progress in this area.

10.3.3 Ultra-Strong Magnets and Accelerators

The starship engine proposals described in preceding chapters require ultra-strong magnets that are not available at present. Superconducting magnet R&D must be strongly pursued to achieve ultra-strong, efficient, magnets for use in starship engines.

10.3.4 Electronic and Materials Technology

The materials used to construct the engines, the structure, and the machinery within the starship should be strong and have a long lifetime. The design life of all components should be at least fifty years.

It would also be desirable to have some of the materials "self-repairing," either in themselves, or through the use of "repair robots."

---

[40] Nuclear submarine reactors usually have a twenty year lifetime.

Here again we have a set of projects to develop super-strong, long life materials as well as machinery and computers that can proceed in parallel with the initial stages of the starship project and be used in the final stages of prototype starship construction.

If we develop multi-stage space gun technology and nuclear rockets at an early stage of the starship construction, then it might be possible to mine and refine materials on the moon to build starships in moon orbit. This would require a sizeable moon colony for mining, refining, and component construction. The moon colony could also become the staging area for starship journeys.

Lastly, we note rockets and starships made of metal are not necessarily required. Other materials may provide a more cost effective approach to the construction of large space vehicles. After all an asteroid is a space vehicle of some endurance although it is not necessarily made of metal.

10.4 Rapid Interstellar Communication

For short starship travel, faster than light communications is not critical. But it is highly desirable.

Recent work on quantum entanglement suggests that instantaneous communication may be possible using this mechanism if an advanced long range form of this laboratory phenomenon can be developed. We describe this concept in more detail in the next chapter.

11. Long Distance Starship Requirements for Travel to Far Stars and Galaxies

If we wish to travel to long distances – up to 100,000 light years initially (roughly throughout the galaxy) – and then millions of light years eventually to other galaxies ("All the Universe"), then critical advances are necessary. A long distance starship program embodying these advances would take place perhaps a hundred or so years after the short distance starship program was completed and after information obtained during journeys to nearby stars is digested.

We see the starship effort as a long term exploration and colonization program in an ever widening ring around earth.

In this chapter we discuss most of the advances that would be needed. Their development will be aided by the R&D effort to build a short distance starship as detailed in preceding chapters.

A major problem of starships is the rapid progress of time on a much faster than light starship. If a starship has a speed that is much faster than the speed of light, then the progress of time in the starship is much faster than the progress of time on earth.[41] For example if the starship is traveling 5,000 times the speed of light, then the increase in time on the starship is 5,000 times the increase in time on earth. In an interval of one year of earth time, 5,000 years will pass on the starship.

The extraordinarily fast passage of time on a very fast starship requires materials, equipment and engines to continue to work effectively for long periods of starship time which is just as real on the starship as earth time is real on earth. **In fact the distance traveled by a**

---

[41] See p. 15 of Blaha (2011c): *All The Universe*.

starship measured in light years is equal to the time of flight to cover that distance measured in years.[42]

Thus we come to the first important long distance starship requirement – long lifetime equipment and starship superstructure. Other requirements follow in this chapter including instantaneous quantum communication.

11.1 Long-Lived Materials

The materials that we use today to build large vehicles such as oil tankers, submarines and aircraft carriers are meant to last up to, at most, a century and often much less. Many of these materials age, deteriorate, rust, migrate within computer chips over time, and actually slowly flow like a liquid in some cases.

Not many materials keep their original characteristics over long periods of time. In the past fifty years there has been much progress in developing new harder, stronger and age resistant materials and metals.

Starships, in which time moves quickly so that thousands and perhaps millions of years of starship time elapse, must be composed of materials with a long stable lifetime. An important part of the R&D for a long distance starship is the development and use of age tolerant materials. The examination of materials used hundreds of thousands of years ago as tools and dwellings shows the ravages of age. A long distance starship should have an initial goal of tens of thousands of years of stability without aging. Ultimately one would hope that starships that don't age in millions of years could be built to travel to other galaxies.

11.2 Long-Lived Machinery and Electronics

If one has materials that preserve their composition, shape and performance characteristics over thousands to millions of years, then one can construct machinery and electronic gear such as computers that can last a similar period of time. Long-lived machinery and

---

[42] Neglecting the time required to accelerate to the starship's speed at the beginning and the time required to decelerate back to a "normal" speed of a few miles per second. See Tables 5.1 and 5.2.

electronic gear then can be used when the long distance starship travels our galaxy and eventually other galaxies.

The intent of most equipment manufacturers is to create products such as automobiles, aircraft and electronic devices with a limited lifetime so that they can sell new products as older products wear out. Long distance starships need materials and machinery that last "nearly forever" – exactly the opposite of the intent of Earth industries.

Consequently a major R&D effort is required to find/develop long-lived materials, and to learn how to construct long-lived starship machinery and electronic gear. This effort should take place during the early stages of the *long distance* starship R&D program so that the needed long-lived materials and components can be used in the latter stages of *long distance* starship design and construction.

11.3 Long Shelf Life Nuclear Reactors and Nuclear Shuttles

Several types of nuclear reactors are required for a long distance starship:

1. A continuous running reactor that can run for up to millions of years to provide power to a starship in flight to a distant location. This reactor may be a low power reactor. It should have a very long lifetime. That this is possible is suggested by the natural nuclear reactor that ran for millions of years about a billion years ago in the Congo.[43]
2. Long shelf life reactors that are not activated until the starship destination is reached. These would power the starship inside solar systems and nuclear shuttles for travel and landings within a solar system.

---

[43] The author suspects that the vast diversity of life in Africa may in part be due to genetic changes caused by this natural reactor. The development of mammalian life may in part also be due to this reactor and the radiation from its waste products and radioactive deposits in the Congo region over the millennia.

11.4 "Instantaneous" Interstellar Communication

Once a starship capability is achieved it will clearly necessitate a very rapid, if not instantaneous, means of communication. All electromagnetic means of communication are limited by the speed of light and are thus insufficient. If neutrinos are tachyons (faster than light) then they could provide a communications channel except that neutrino detection is very difficult, not reliable, and would require massive detectors that would be an unacceptable addition to the mass of a starship.

The only possible method appears to be based on quantum entanglement – currently a subject of intense scientific interest. Based on current thinking about this form of quantum communication it will have the following very desirable features:

1. It is a 1:1 form of communication with no possibility of being intercepted by others.
2. It requires a small amount of power no matter what the distance.
3. It is instantaneous and thus gives direct real time communications over any distance – even millions of light years.

If history is any guide the development of interstellar communications will be similar to the development of telecommunications over the past 150 years. Thus we anticipate that it will begin with a primitive Morse code equivalent, and progress eventually to fast digital transmission of images and data. We anticipate bilateral switchboards initially that eventually will lead to communications networks among a growing group of star system colonies. Obviously this capability would be needed for exploration – particularly by robot driven starships, and for communications between colonies.

The basic mechanism module will consist of a bilateral quantum entanglement setup that begins as two electrons[44] of opposite spins in a quantum state with total spin zero. Each electron is nudged into a magnetic bottle that does not affect their joint spin state. One bottle is

---

[44] Protons would be another reasonable alternative.

retained on earth; the other bottle is placed in a starship. As the starship travels the state of the electron spin within its bottle can be periodically sampled but without changing its state.[45] This can also be done on the earth based electron in its bottle. If either electron's spin is flipped the spin of the other electron flips instantaneously no matter what the distance. Thus instantaneous communication of one computer bit takes place. Eight such bottle pairs allow us by flipping bits to exchange bytes of data. Because of the time contraction associated with faster than light starships the byte change must be almost instantaneous for effective communication between a starship and earth. Between star colonies whose relative motion is small the byte "exchange" need not be "instantaneous."

Eventually arrays of bottles can transmit bytes in bulk in support of large data and image transfer. One can envision electronic switchboards eventually linking arrays of bottles to form a network. The thought processes and designs are similar to those in telecommunications.

One might ask if instantaneous quantum data transfer is possible. Both quantum theory and numerous experiments have shown that instantaneous data transfer via entangled pairs works at large distances.[46]

11.4.1 Interstellar Communications and SETI

If our (and others) suggestion that quantum communication is the only reasonable way for communications at large distances, then this might be the reason for SETI's failure to find communications by alien civilizations. They may very well not be communicating by radio or laser waves.

It is important to note that quantum communication, as we have proposed it, is inherently private 1:1 communication with no visible manifestations for others to detect of which we are aware. Also quantum communications throughout the universe do not need

---

[45] 2012 Nobel Prize winner Serge Haroche of France developed ways of detecting the state of particles without disturbing their quantum state.

[46] Matson, John, Quantum Teleportation Achieved Over Record Distances, Nature, 13 August 2012.

powerful transmitters realizing a power saving on starships. So signals of massive electromagnetic power usage would not be seen.

11.5 Suspended Animation for Long Trips

As we noted earlier it would be helpful to have suspended animation available for crews on short distance starship journeys. In the case of long distance starship flights suspended animation is absolutely necessary for the crews. With suspended animation a crew could go on a journey lasting millions of years of starship time, and, upon return to earth, have aged physiologically only a short time while out of suspended animation exploring distant star systems. The round trip travel time will not have aged them. When they return to earth they may be some months older, but their families and friends (having aged only by the earth travel time) will still be roughly contemporary with them.

A mechanism for long term suspended animation is thus a major requirement. Any suspended animation mechanism must take account of three important facts: 1) suspended animation must reduce human body temperatures to a low value to "halt" life processes and bodily decay; 2) lowering body temperatures will cause cells to rupture due to the expansion of water upon freezing; 3) the entry into suspended animation and the reentry to a normal bodily state must be rapid and uniform throughout the body.[47]

A mechanism to achieve these goals is not presently known. The current approaches to suspended animation (which all include lowering body temperature) are:

1. Replacing part or all of the blood in an organism with an "antifreeze" solution that will prevent cells and body tissue from bursting when the temperature is lowered. Revival takes place by raising the temperature of the organism while returning blood to the organism's circulatory system. This approach has been successfully applied to dogs that have been put into

---

[47] One cannot "unfreeze" part of a human body and have the rest still frozen.

suspended animation for three hours. Unfortunately some of the dogs had nerve and coordination problems after revival.[48]

2. An organism can have a chemical injected or absorb a chemical while breathing that will counteract the tendency of water to expand when body temperature is lowered and/or lower the metabolic rate of the organism.

3. NASA and other groups have studied the possibility of placing humans into hibernation. Since hibernating organisms do age – perhaps more slowly – this approach is not true suspended animation.

While we do not know of a reliable and effective method for suspended animation an ideal general method of suspended animation is:

Entering Suspended Animation
1. Establish non-expanding water in body (Possible mechanism: Destroy the appearance of crystalline structure in the transition of water to ice. Ice is less dense than liquid water due to a crystalline structuring of ice. The crystalline structuring might be preventable with electromagnetic radiation in some form.)

2. Instantly freeze the person throughout his/her body with a "reverse" microwave mechanism that totally freezes all parts of the body simultaneously. Such a heat withdrawal mechanism is possible thermodynamically.

Restoring a Person to Normal State
1. Instantly restore 98.6 degree normal body temperature throughout body by a "microwave" heating mechanism.

2. "Electroshock" the body to turn on "all" bodily functions immediately.

---

[48] At the University of Pittsburgh's Safar Center for Resuscitation Research.

Clearly current approaches are not anywhere near a form of suspended animation that long distance starship journeys require. The biomedical study of forms of suspended animation is underway and should receive strong support as part of the starship programs and for medical benefits that would also result.

11.6 Robotic Driven Starships

The initial starship flights, especially the long distance flights, could be manned by robots rather than humans. This approach would be useful to test the stability and longevity of the starships, and their components, without endangering a crew. The robot guidance systems would, of course, have to be constructed of long-lived components. If they are successful then a robotic trip would also help demonstrate the long term reliability of long-lived computer equipment.

Robotic flights would be especially useful if a method of rapid faster than light, or instantaneous, communication between the starship and earth existed. This possibility has been explored above. A sufficiently sophisticated communication capability might go so far as to transmit pictures of worlds around other stars as well as scientific data. In addition, changes in flight plan could be sent from earth – a useful capability since some remote planets will be more interesting than others – particularly earth-like planets. The possibility is also present for intercepting and sending alien electromagnetic messages local to a destination solar system.

11.7 Long-Life Computer Chips

Computers hardened for battle and bad weather conditions currently exist. A long distance starship would require computers with working lifetimes of between thousands and millions of years. In time periods of that length the chips within the computers would be subject to aging processes such as the intermixing of the metals composing the various chips of the computer and the aging of the wiring of the computer. Since new materials of greater strength and other superior properties are being discovered fairly frequently one can hope that the required types of metals and materials will be found.

Thus a need exists to develop computers that can perform tasks correctly for extremely long periods of time. This appears to be a challenging R&D task that could be initiated during the short distance starship R&D program given sufficient funding.

11.8 Space Dust

The effect of dust and gas molecules in space on the starship is of great importance. This effect should be detectable in short distance starship voyages. If it is important, as it seems to be, then the design of the shielding for long distance starships should incorporate appropriate "armor" to protect the starship and crew.

11.9 Length Dilation Effects

Lengths on a starship, traveling at high speed much greater than the speed of light, appear to be dilated - longer. A length measured on a starship will appear to be larger to an observer on earth by a factor of the speed measured in terms of the speed of light than the length on earth. For example if a starship is moving at 5,000 times the speed of light then a two meter long stick (earth length) would appear to be 10,000 meters long on the starship to an earth astronomer viewing the stick on the starship.[49]

Does this length dilation phenomenon affect the contents of the speeding starship? No. It is an illusion that the earth observer "sees." But an occupant of the starship would not notice a difference and would see the stick as still two meters long.

---

[49] This discussion assumes that the stick and the starship motion are parallel. For a detailed discussion of this length contraction phenomena see Blaha (2011c) p. 17.

12. Time Frames for Research and Development

The United States, China and Russia would be individually capable of implementing the short range starship program. However it would be more satisfactory for these three nations and other groups such as the European Union to form a consortium to utilize the scientific and engineering resources of all these nations, and also to share the costs.

12.1 Multi-Stage Space Guns

The development of multi-stage space guns would appear to take up to ten years. The R&D procedure should, in part, emulate Japanese efforts to reduce the percentage of defects in computer chips. In the 1980's Japan's chip industry built and rebuilt chip manufacturing devices noting causes of defects in the chip fabrication process. By continually, incrementally improving the process they were able to develop chip manufacturing devices that had very low defect rates.

Similarly a series of prototypes of the multi-stage gun with a detailed analysis of results for each prototype should lead to a reliable, cost-effective space gun that can loft 500 kg payloads into space. An array of 100 such guns spread over a level 150 mile by 150 mile part[50] of the southwest US desert could send 50,000 kg per day into Low Earth orbit with the help of scooper rockets in space that would capture payloads and place them in the required orbit.

The construction of the 100 guns and support facilities would take at least 5 – 7 years. The cost of this effort from the beginning to the placement of the guns would be of the order of a few billion dollars.

---

[50] The space guns would be separated from each other by 15 miles to minimize vibrations from affecting each other and to avoid possible collisions in space. The guns would be shot on a schedule - perhaps 4 shots per day per gun. Between shots the guns would be inspected and repaired if necessary as well as being loaded for the next shot.

An enormous increase in our ability to send material into space would result to the benefit of a large space station, a moon base and perhaps a Mars colony. Man would be able to transcend earth's gravity.

12.2 Nuclear Rockets

Nuclear rockets for use in our solar system can be built rather quickly based on the prototype HARP designed earlier. They are thermal nuclear rockets. The design of its nuclear reactor(s) can be based on the long experience with nuclear reactors for power plants.

The major part of the cost is transportation of the rocket components into a Far Earth orbit for reasons of safety in case of a mishap. It appears that the cost of designing and then building the nuclear rocket in space would be several billion dollars and take up to ten years. (The Russians estimate the construction of a nuclear rocket on earth would cost about 600 million dollars.) The transportation of nuclear rocket components to space need not wait for multi-stage space guns to become available.

The R&D of Pulsed Pellet Micro-Nuclear Explosion Rockets is a significant project. It will take upwards of ten years to create a working prototype on earth. The cost would be of the order of a billion dollars. Constructing this type of rocket in space and its testing would also cost several billion dollars unless multi-stage space guns can be used to significantly lower transportation costs.

Long shelf life nuclear rockets also require a significant R&D effort. It would probably take ten to 15 years to develop safe, reliable prototypes and cost several billion dollars. The construction of these types of rockets in space and their testing would also probably cost several billion dollars unless multi-stage space guns substantially lower the cost of transportation to space.

12.3 Short Distance Starship

Section 10.1 describes the various R&D stages, and estimates a total cost of 617 billion dollars and a total starship construction time of 58 years.

12.4 Long Distance Starship

It is difficult to estimate the construction time and cost of a long distance starship due to the critical requirements specified in chapter 11 and earlier chapters. However, in view of section 12.3's estimates and assuming we can build on the short distance starship R&D it seems that it would require perhaps 50 additional years after the short range starship and perhaps an addition 250 billion dollars. Thus we arrive at the estimate of a total time to delivery of 108 years[51] and a total cost of 867 billion dollars.

Much of this cost and construction time would be expended on the short distance starship although the long distance starship would have to be separately constructed in space. So the estimates above embody some double counting. A bold approach, at the construction stage, would be to build both starship superstructures in parallel with some attendant cost savings.

---

[51] NASA has estimated that a starship would take 100 years to construct. It is not clear whether they are referring to a short distance or long distance starship.

13. Needed: Earth Exploration

To understand alien life, intelligent or not, should we encounter it in space we must first understand the variety of life on earth. It is clear from the almost weekly discoveries of new species and forms of life both in extreme environments such as underwater volcanic vents or deep within the earth that there is much to learn. (Most of the living matter on earth is significantly below the earth's surface to depths of the order of 3 km.) When we explore Mars as we are doing now and note that it is likely most of its water is below its surface, then we must strongly consider the possibility that Martian life is subterranean – avoiding the harmful radiation on its surface.

So, as a preliminary to considering life on other worlds, we will briefly consider life on earth – a subject currently under widespread investigation.

In addition in appendix A we consider the effect on world civilizations if a more advanced alien species is encountered and has a continuing intercourse with earth civilizations. The results are based on our theory of civilizations that has successfully explained the dynamics of over 50 earth civilizations. We find the impact of an alien civilization will generally take hundreds of years to affect earth civilizations beyond the level of scientific gadgets and physical theories.

13.1 Some Unusual and Extreme Environments on Earth

Many projects are underway to find new species and forms of life on earth. These regions include areas occupied by humanity (on the earth's surface, in water, in mines and caves, and so on) but still having undiscovered forms of life. They also include the Amazon basin, Africa, India, Tibet, Indonesia, Southeast Asia, New Guinea, Australia, and Antarctica. The types of regions include deserts, polar regions, ocean depths, volcanoes – above and below sea, deep probes within the earth, upper air bacteria/microbes in the air and in clouds. Very exciting efforts are being made to explore for life in deep lakes under domes of ice in Antarctica. These lakes have been isolated from the earth's surface for up to millions of years.

An understanding of novel life forms on earth, and their biology, will help us understand any new forms of life that may be found on other planets and moons (and perhaps floating in space on small bodies of matter or free "floating" in space.)

These explorations of earth life nicely complement preparations for extended space exploration.

13.2 Semi-Intelligent/Intelligent Species on Earth and Ocean

The earth has many species that might be called semi-intelligent or intelligent in a different way from human intelligence. (Human intelligence also has widely varying types buried within the mass of humanity. There is no rigorous method of differentiating between the types of human intelligence. So the types are at best differentiated by labels: genius, idiot, moron, and so on.)

We are aware that many intelligent and semi-intelligent species exist on earth. Many of them are in the process of extinction such as the great apes and whales. A short list of species exhibiting at least semi-intelligence is: apes, monkeys, elephants, some bird species, dolphins, porpoises, whales, and some reptilian species such as crocodiles and alligators.[52]

There are several points to be made with respect to these species:

1. Many of them are being killed off by Man.

2. Despite our presumed superior intelligence we have not been able to understand or speak anything more than simple aspects of their forms of communication.

---

[52] One could add dogs and other species to this list since they exhibit intelligence in certain situations. Those who feel crocodiles and alligators are not in any way intelligent should consider the case of a man in Florida who was hunted by an alligator for several days before being killed by it.

3. Some of these species have attempted to speak human language in some fashion. In at least one case, not having our vocal cords, they have attempted to simulate human speech using breathing techniques.

4. Some species are known to be able to count and do simple arithmetic such as determining which pile of food is larger.

5. Many species show human-like emotions such as anger, love, loyalty, and concern for others. Some, such as elephants, display family and social feelings.

The failure to communicate with other species with which we have had long contact in a common environment suggests that Man will have difficulty communicating with any alien species that it encounters unless the alien species is so much more advanced than Man that it can learn a language of Mankind. Mankind would be the junior partner in this dialogue.

13.3 Social Insects

In the preceding section we have not mentioned insects as possible intelligent or semi-intelligent species. Non-social insects do not appear to have intelligence but seem to be governed solely by instinct.

However some species of social insects: bees, wasps and ants do appear to have social characteristics that could be interpreted as at least semi-intelligent. One study suggested that a swarm of bees through various forms of motion (gestures and dances) appeared to act intelligently in certain situations.

Again we are faced with our inability to understand a form of communication. At best we recognize a possibility of intelligent behavior. Thus we see that encounters with alien species will present major communications difficulties.

13.4 Networked Microbes and Plant Life

On earth we have come to believe that only animal life can have a form of intelligence. However that belief may be in a process of change. First it has long been noted that plant life often "arranges itself" for maximal benefit. A simple example is the tendency of some trees to space themselves so that they will not form an overly dense thicket that would prevent normal growth. More subtle forms of plant communication, that can be called semi-intelligent or not, also have been recently found. For example there is a species of microbes that live in soil that extend long fibers that the microbes use to exchange information about soil and water conditions. Also some plants exude chemicals that enable communications between them.

These simple examples on our animal dominated world may be harbingers of intelligent plants with significant communications capabilities amongst their species on other worlds. Thus we see the possibilities of semi-intelligent and intelligent species on other worlds are vast. The problem is communication judging from our experience on earth.

Appendix A. Interaction of Earth Civilization with an Alien Civilization

The bulk of technical civilizations in the universe may be immensely more advanced than ours ... enormous, almost unbelievable, quantities of information can be communicated over immense distances, if such civilizations exist.
Carl Sagan

This appendix is mathematical in nature. Those not mathematically inclined can skip this appendix.

In a series of books by this author on civilizations we treated the possible consequences of the encounter of earth civilizations with a more advanced alien civilization. (The case of an encounter with a less advanced civilization was not considered because we would then play the role of "teacher" and earth civilizations would at most be affected culturally. However given the wide cultural variety of earth civilizations encounter with an alien, less advanced, culture would not appear to significantly affect the growth of earth civilizations.)

The transfer of science and technology from the alien civilization to earth civilizations would appear to stimulate growth in earth civilizations in a manner similar to the Industrial Revolution with major growth in all measures of a civilization. It would also be similar to the impact of the migration of Greek scholars to Italy: the Renaissance.

Using the growth equations that we have developed in Blaha (2010c), *SuperCivilizations: Civilizations as Superorganisms*, and earlier books we see that the growth is not immediate but rather takes place over periods of hundreds of years. The growth period depends on the relative difference between earth civilizations and the alien civilization.

We reprint chapter 22 of Blaha (2010c) below to give the reader an idea of the impact of the discovery of an advanced extraterrestrial civilization.[53] We assume the contact will be friendly and mutually advantageous.

A.1 Extraterrestrial Civilizations

The search for extraterrestrial life has not as yet uncovered evidence of life "out there." But recently evidence has appeared for many of the preconditions of life as we know it. Water has been found in large quantities on Mars as well as on some Jovian satellites (which may have extensive liquid water oceans beneath their surfaces.) Water and organic compounds have also been detected in interstellar regions suggesting that the preconditions for life may be fairly widespread in the universe. Lastly, life seems to have appeared on earth about five hundred millions years after it was formed – rather quickly from the point of view of astronomical times. These considerations suggest extraterrestrial life will eventually be found (and perhaps very soon if the proposed Mars expeditions are successful.)

A.1.1 New Searches for Extraterrestrial Life

Several other major initiatives will start in the next eight years to identify earth-like planets within 4,000 light years of the earth. The National Aeronautics and Space Administration will launch spacecraft that will systematically search for earth-like planets: the Kepler Spacecraft Mission and the Space Interferometer Mission - SIM. The Kepler mission will observe up to 100,000 stars over a period of two to three years. It will be capable of detecting earth-sized planets circling stars up to a distance of 4,000 light years away by sensing the dimming of a star's light as an earth-sized planet rotates around it casting a shadow. After this survey is completed we will have an idea of the number of earth-like planets in the galaxy. We will also have an idea of where to point antennas to attempt to receive radio and/or laser transmissions from these planets.

---

[53] Societal Level is an indicator of the general level of the civilization: culture, technology, social structure, cohesion and so on.

The Space Interferometer Mission (SIM) will start measuring the wobbles of nearby stars within a radius of 50 light years of earth. It will be sensitive enough to detect earth-sized planets that are orbiting a star at a distance of 0.5 AU to 10 AU from the star. (An AU is about 93,000,000 miles – the distance of the earth from the sun.)

A third spacecraft, the Terrestrial Planet Finder (TPF), is scheduled to be launched around 2015. Its goal is to chemically analyze the atmospheres of promising planets found in preceding space flights to search for evidence of life. The TPF will gather light reflected by a candidate planet and analyze the light for evidence of water, carbon dioxide, oxygen or methane in the planet's atmosphere. The presence of these chemicals would suggest the possibility of life on the planet.

The net result of these three missions will be to identify the extent of life in the galaxy and to identify earth-like planets in the general vicinity of earth. With this knowledge it would be possible to target likely planets that might harbor intelligent life and then attempt to detect radio and laser transmissions from these planets. Ultimately, conventional probes could be sent to nearby planets to explore them and perhaps return. These probes would take hundreds of years to reach planets circling nearby stars.

A.1.2 Communicating with Extraterrestrial Civilizations

After extraterrestrial life is found the next step will be the detection of extraterrestrial civilizations and the establishment of communication with them. The imaginative Project SETI represents an inspiring example of efforts in this area—particularly with its exciting use of ordinary individual's personal computers to analyze data for signals. When evidence for extraterrestrial civilizations is found there will still be the formidable technical challenge of establishing a practical form of communication with them. (Beyond that is the problem of fast interstellar travel for communication, trade and exploration discussed earlier.) The apparent limiting factor of the speed of light is the major impediment to effective communication. Ordinary radio and TV-style communication would take years to communicate with potential nearby civilizations.

There have been suggestions that the limitation of the speed of light might be evaded through the use of quantum effects. Some buzzwords in this area of study are quantum teleportation, quantum communications and quantum entanglement. Such efforts are currently highly speculative.

In this chapter we will extend our study of civilizations to include extraterrestrial civilizations. We will base our analysis on a common assumption behind much of the work in the biological and social sciences: that our knowledge of current times and nearby places can be extended to other times and other places. Nature uses similar methods everywhere and no place has a special significance or special treatment by nature. This assumption is true in the natural sciences and would seem to be true of social institutions as well since they are the result of natural events.

With this as our starting assumption we observe that the most intelligent species on earth tend to have long lifetimes. We think of man or elephants or whales with lifetimes that can be over a hundred years. In addition the more intelligent species tend to have a longer period of adolescence before they become mature adults. We also note that as mankind has advanced the average lifetime of mankind has increased significantly. For these reasons it seems reasonable to assume that civilizations that are much more advanced than earthly civilizations have populations with much longer life spans. The period for routs and rallies of earthly civilizations seems to be eight generations or approximately 267 years. The periods of alien civilizations could be much longer.

A.2 Forms of Extraterrestrial Civilizations

An extraterrestrial civilization at about the same level as Western Technic civilization might thus have a population with a similar lifetime as humans and perhaps a similar period for routs and rallies.

A very advanced extraterrestrial civilization might have a much longer average lifetime for their population and a much longer period of oscillation. Therefore we will look at the Societal Levels of civilizations with periods of 800, 8,000 and 20,000 years. The choice of these periods was motivated by simple considerations:

1. The choice of a period of 800 years was based on the concept of a civilization that had a period roughly three times longer than earthly civilizations. The average lifetime of a person in Western civilization has doubled in the last century. Recent medical advances suggest that a significantly longer lifetime for humans is feasible in the near future. If humanity's average lifetime triples with a corresponding tripling of the length of generations then an 800 year oscillation in human civilizations may be attainable in the future. An extraterrestrial civilization with an 800 year period thus seems reasonable as well.

2. The choice of an 8,000 year period was simply an extrapolation by a factor of ten. It seemed suitable for a very advanced civilization. Earth has species (trees and shrubs) with lifetimes of thousands of years. So the possibility of intelligent creatures with very long lifetimes is worth considering.

3. The choice of a 20,000 year period was motivated by a curiosity to see how a civilization with a universal state of 30,000 years duration would look. The motivation was to compare it to the Galactic Empire of Isaac Asimov's Foundation series of books. Asimov's Empire had a 30,000 year lifetime. It appears Asimov's Empire is not consistent with the theory of civilizations presented herein. The population of Asimov's Empire has the same life span and social characteristics as the populations of current earth civilizations. Therefore a 30,000 year lifetime for a universal state is very unlikely. In addition, the discussions of the nature of the Empire, its decline, and the following barbarous period also differ from our theory and Toynbee's (and other historians') observations.

A.2.1 Setting the Parameters of Extraterrestrial Civilizations

The period T sets the parameter b of our theory of civilizations directly due to equation (16) of subsection 2.1.2:

$$T = 2\pi/b \qquad\qquad (A.1)$$

The following table lists values of b and the lengths of the phases of a civilization for each choice of period. The generation length is based on eight generations per period as seen on earth. It is only meant to convey an idea of time scales.

Period (years)	Generation Length (years)	b (years^{-1})	Startup (years)	Time of Troubles (years)	Universal State (years)	Total Lifetime (years)
267	33.38	.0235	134	400	400	934
800	100	.00785	400	1,200	1,200	2,800
8,000	1000	.000785	4,000	12,000	12,000	28,000
20,000	2500	.0000785	10,000	30,000	30,000	70,000

Figure A.1. Possible time periods of extraterrestrial civilizations.

The other parameter that describes a civilization is the parameter a. The value of a sets the rate of the decline of a civilization. While some might hope that a sufficiently advanced civilization might not decline like earthly civilizations, the decline of all complex entities with time is a universal fact of nature. So we shall assume that extraterrestrial civilizations will also decline with time. The rapidity of the decline is not something that we can at present determine. Therefore we will consider a range of possibilities based on earthly experience extrapolated (adapted) to extraterrestrial civilization time scales.

The successive peaks of a civilization decline by the *factor*:

$$e^{-d}$$

where

$$d = aT \qquad\qquad (A.2)$$

with T being the period measured in years by equations (18) and (19).

For human civilizations we found d = .75 and thus a = .00281 were able to describe all the human civilizations that we studied except for petrified civilizations where d = 5 (and thus a = .0187) was used. For extraterrestrial civilizations we will consider a range of d values from .5 to 5. The values of the parameter, a, that we examined are:

$d =$	.5	.75	1	2	5
$e^{-d} =$	.607	.472	.368	.135	.00674
Period					
267	.00187	.00281	.00375	.00749	.0187
800	0.000625	0.000938	0.00125	0.0025	0.00625
8,000	0.0000625	9.38E–05	0.000125	0.00025	0.000625
20,000	0.000025	3.75E–05	0.00005	0.0001	0.00025

Figure A.2. Possible values of the variable a.

Using the above values of a and b, and the universal expression for S given in equation (13) with $c_1 = 1$:

$$S(t, r) = [b - e^{-at}(a \sin(bt) + b \cos(bt))]/(a^2 + b^2) \qquad (A.3)$$

we obtain the following figures displaying the growth/Societal Level S for extraterrestrial civilizations.

A.3 Extraterrestrial Civilization Interacting with a Human Civilization

The interaction between an extraterrestrial civilization and an earth civilization such as Western Technic civilization is an interesting question to investigate. The first issue to address is how can an earth civilization interact with an extraterrestrial civilization. There does not appear to be any extraterrestrial civilizations in this solar system although the possibility of hidden civilizations beneath the surface of Mars has not been ruled out. There is also a possibility of life on Jupiter and Saturn's satellites that might have evolved into intelligent species with civilizations. But all these possibilities are unlikely. We have not detected any electromagnetic radiation from these bodies that would signal a civilization. So if a civilization exists within our solar system it must either be primitive or completely different in nature from our technological civilization.

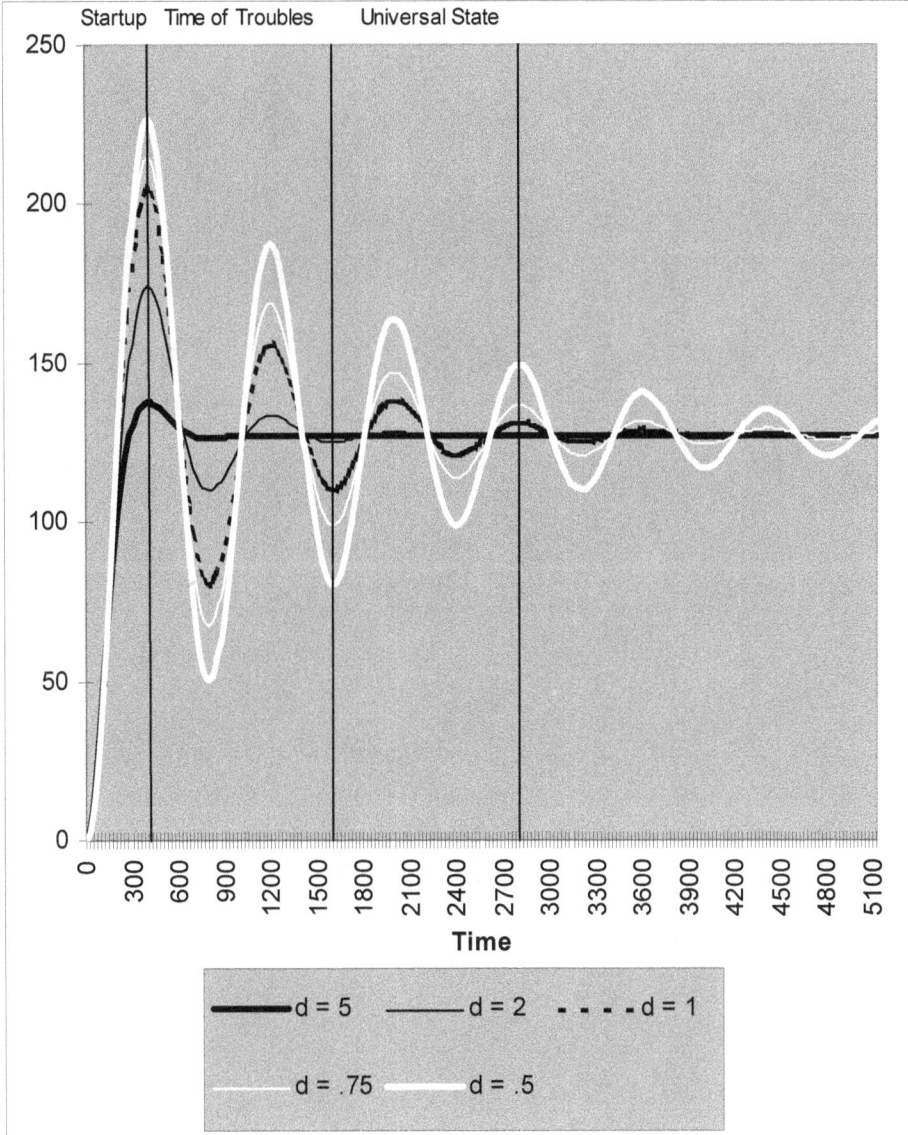

Figure A.3. S(t) for extraterrestrial civilizations with an 800 year period.

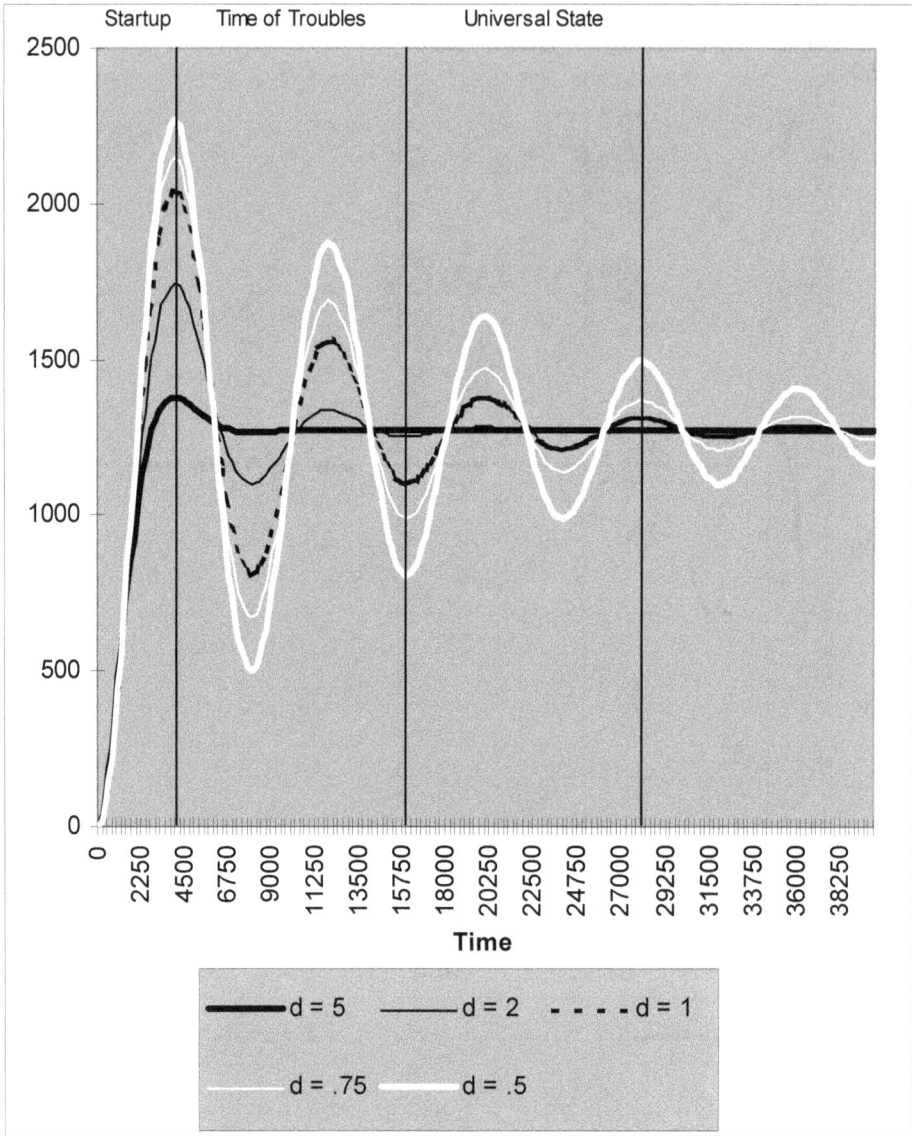

Figure A.4. S(t) for extraterrestrial civilizations with an 8,000 year period.

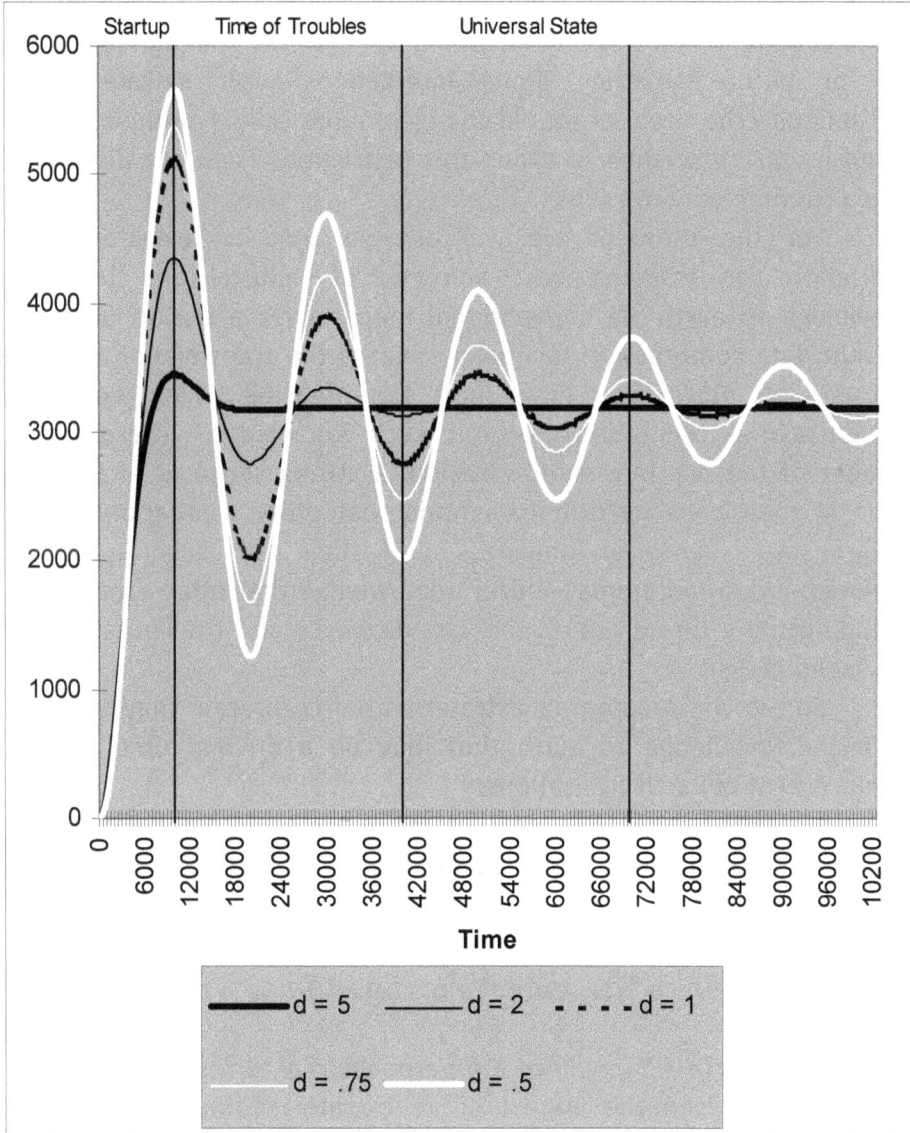

Figure A.5. S(t) for extraterrestrial civilizations with a 20,000 year period.

Thus we expect that any extraterrestrial civilization must be located in another solar system(s) in our galaxy or in other galaxies. Interaction with a civilization at these great distances will be difficult due to the limitations on communication and travel imposed by the speed of light. No message can travel faster than the speed of light. This law of nature severely limits interactions with extraterrestrial civilizations. (The situation would change dramatically if methods can be found by future science to evade this restriction. Then our discussion would become very relevant.)

For the moment we will assume that an extraterrestrial civilization can transmit vast amounts of cultural and technical knowledge to earth via conventional means over a period of years. Current data compression technology makes the transmission of large amounts of information in a short time feasible. Why would an extraterrestrial civilization transmit such knowledge? There are a number of reasons one could envision: a strong sense of altruism, a pride in their own civilization, a hope that other civilizations might reciprocate, a sense of loneliness after their own solar system is explored and found empty, and/or a knowledge that other civilizations would not be a threat due to the vast distances and the limitations of the speed of light.

So we will assume an extraterrestrial civilization may transmit advanced knowledge to earth that may be a driving force for the advancement of earth civilizations.

We will further assume that the extraterrestrial civilization is described by equation (13) using extraterrestrial values for a and b which we will denote as a_e and b_e. (e is for extraterrestrial.)

$$S_e(t, r) = c_1[b_e - e^{-a_e t}(a_e \sin(b_e t) + b_e \cos(b_e t))]/(a_e^2 + b_e^2) \qquad (A.4)$$

with $c_1 = 1$. (Actually S_e could be a superposition of terms.)

We will see the impact of an extraterrestrial civilization on Western Technic civilization in the effects of the extraterrestrial civilization's force. (We assume the interaction will be peaceful and constructive. The possibility of the obliteration of human civilizations also exists.) We will represent the force exerted by the extraterrestrial civilization on a human civilization starting at time $t = t_0$ by:

$$F_{etext} = \gamma \, S_e(t) \, \theta(t - t_0) \tag{A.5}$$

where γ is a constant. The time t_0 is the time of the first significant exchanges between the extraterrestrial and the human civilization.

The solution of

$$mC'' + rC' + sC = F_{ext} + \gamma \, S_e(t) \, \theta(t - t_0) \tag{A.6}$$

(with F_{ext} containing all other forces) leads to

$$S = S_0(t) + \gamma b_e(\, t - t_0)/(mD) + S_1(t) \tag{A.7}$$

$$S_1(t) = -\gamma b_e[(s(t) - s(t_0)]/m \tag{A.8}$$

with

$$s(t) = e^{-a_e t}[A_e cos(b_e t) - B_e sin(b_e t)]/ (b_e(b_e^2 + a_e^2)^2 E) + \\ + e^{-at}[A cos(bt) - B sin(bt)]/ (b(a^2 + b^2)^2 E) \tag{A.9}$$

where

$$A_e = 4a_e b_e^3 - 2ab_e^3 - 2a_e^2 b_e(a_e - a) - 2a_e b_e[b^2 + (a_e - a)^2] \tag{A.10}$$

$$A = 4ab^3 - 2a_e b^3 - 2a^2 b(a - a_e) - 2ab[b_e^2 + (a_e - a)^2] \tag{A.11}$$

$$B_e = b_e^2(b_e^2 - b^2) + a_e^2(a_e - a)^2 - 6a_e b_e^2(a_e - a) - a^2 b_e^2 + a_e^2 b^2 \tag{A.12}$$

$$B = b^2(b^2 - b_e^2) + a^2(a_e - a)^2 + 6ab^2(a_e - a) + a^2 b_e^2 - a_e^2 b^2 \tag{A.13}$$

$$D = (a_e^2 + b_e^2)(a^2 + b^2) \tag{A.14}$$

$$E = [(b_e^2 - b^2)^2 + (a_e - a)^4 + 2(b_e^2 + b^2)(a_e - a)^2] \tag{A.15}$$

where $S_0(t)$ is the Western Technic Societal Level in the absence of the extraterrestrial force, and a and b are the parameters describing an earthly civilization. This solution applies unless $a_e = a$ and $b_e = b$. (In the unlikely case that the extraterrestrial civilization has the same a and b values as earth civilizations the solution for the barbarian - civilization interaction would apply with the earth civilization being the "barbarians".)

The parameters a and b are defined as earlier. We use our standard terrestrial values for a and b: a = .00281 and b = .0235. We will consider the cases of an extraterrestrial civilization with a period of 800 years and an extraterrestrial civilization with a period of 8,000 years. In each case we will use d = .75 to set a_e (see the previous section). The value of b_e is set by the period of the civilization as in the previous section.

Period (years)	a_e	b_e
800	.000938	.00785
8000	.0000938	.000785

We will choose the time that serious interactions begin in the not too distant future: t_0 = 2038. It is conceivable that Project SETI or other efforts will result in contact within the next thirty-six years.

The interaction constant γ should depend on the earth civilization's receptivity to transmissions from other cultures. While γ could be an arbitrary constant that differs in an unpredictable way from interaction to interaction, we will assume that it depends on the receiving civilization's internal characteristics. In particular we set

$$\gamma = 4mb^4 \qquad\qquad (A.16)$$

where b is the b value of the receiving earth civilization in this section. We shall see that this choice leads to what appears to be reasonable results for both the 800 year and 8,000 year extraterrestrial civilization cases. Figures A.6 and A.7 show the plots of eq. A.7 for each case.

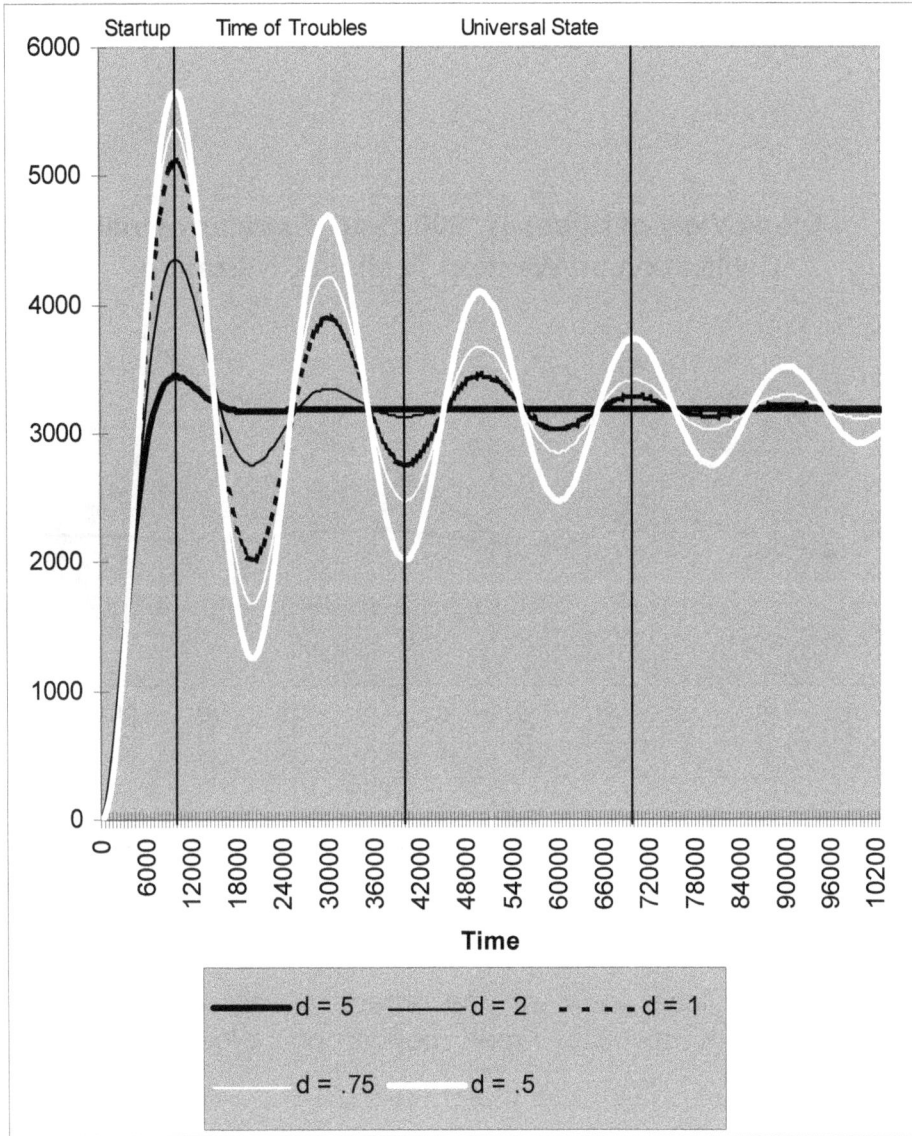

Figure A.6. Effect of an extraterrestrial civilization with an 800 year period on Western Technic civilization.

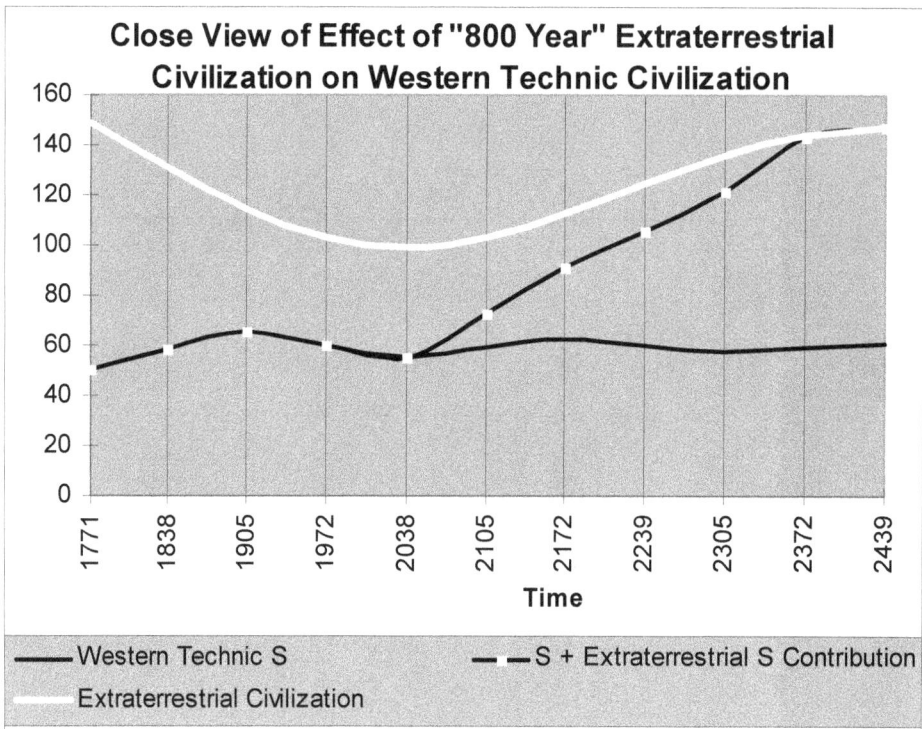

Figure A.7. Close view of the effect of an extraterrestrial civilization with an 800 year period on Western Technic civilization.

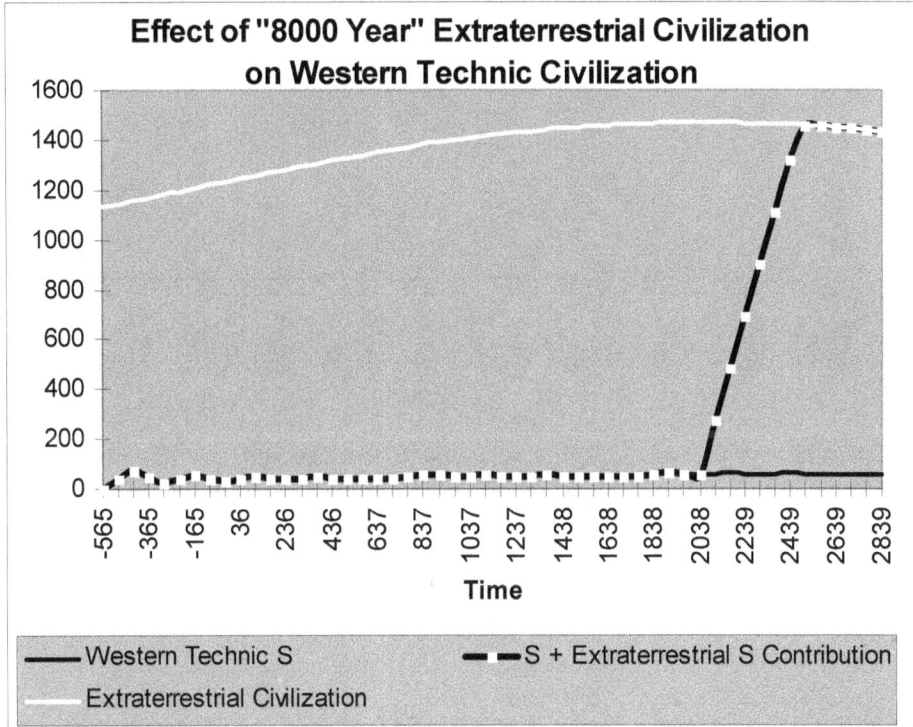

Figure A.8. The effect of an extraterrestrial civilization with an 8,000 year period on Western Technic civilization.

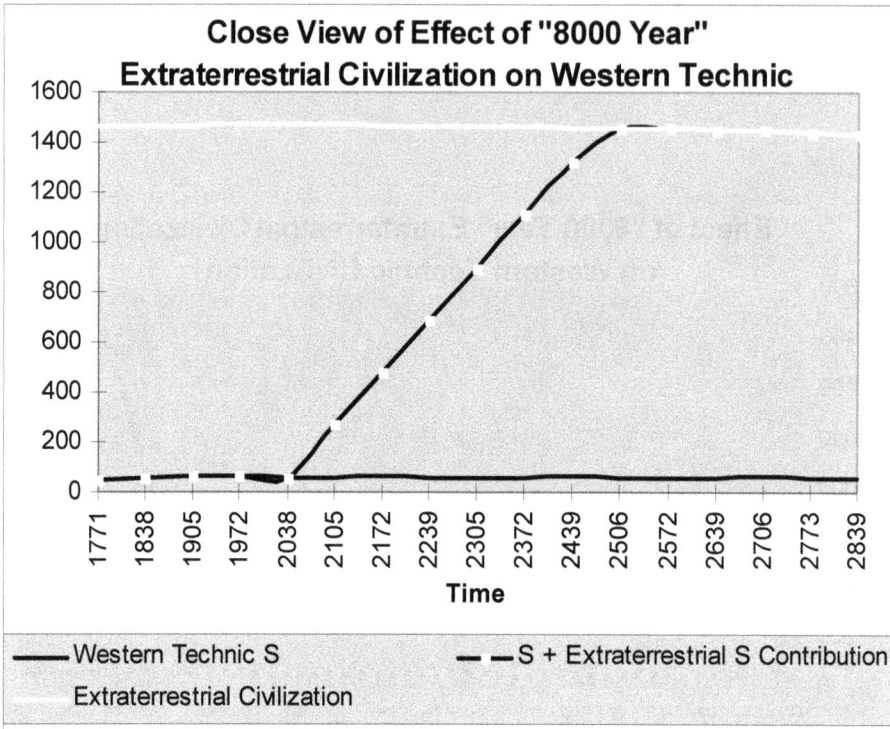

Figure A.9. A close view of the effect of an extraterrestrial civilization with an 8,000 year period on Western Technic civilization.

A.4 The Resulting Rapid Rise of Human Civilizations

If an advanced extraterrestrial civilization made contact with earth civilization, then the preceding figures suggest that earth civilizations would progress upwards relatively rapidly to the level of the extraterrestrial civilization. The form of contact could be as simple as a one-way transmission of the scientific and cultural knowledge of the civilization via radio waves or laser beams. Earth scientists could decipher the incoming data via some "Rosetta stone" perhaps in the form of mathematics. The following table shows the transition time for technical earth civilization to reach the level of advanced extraterrestrial civilization is of the order of a few hundreds of years.

Extraterrestrial Civilization	Time Required By Earth Civilization To Reach its Level (years)
800 Year Period	334
8,000 Year Period	468

Figure A.10. The time required by earth civilization to reach the level of various extraterrestrial civilizations in our theory.

Although the time required for the transition is not long on the scale of human history it can expect to be punctuated by massive civil disturbances unless carefully managed.

A.5 Dominant Term in Societal Level Transition

The dominant term in the transition expression for S given by eq. A.7 is the linear term in time:

$$\gamma b_e(t - t_0)/(mD)$$

which emanates from a constant term in the change C(t) that, in turn, has its source in the constant term in the force exerted by the extraterrestrial civilization on the earth civilization.

Consequently any extraterrestrial civilization that has a roughly constant growth/Societal Level during the period of contact with earth civilization can be expected to generate such a linear term in the growth/Societal Level of the earth civilization. Thus our results have some generality and do not depend on the precise form of the extraterrestrial civilization's growth/societal curve S(t).

A.6 Stimulus via an Extraterrestrial Information Burst

The force exerted by the extraterrestrial civilization may take place in a short time interval: for example in the form of a massive burst of information sent to earth over a period of days or weeks, and containing the "entire" accumulation of knowledge of the extraterrestrial civilization. There are two extremes of response that one can envision:

1. This burst of data could have the same overwhelming effect on Western Technic civilization that the Hyksos conquest had on Egyptaic civilization: but as a conquest of knowledge and ideas rather than a physical, military conquest. The result could be a revulsion against the influx of knowledge (while perhaps accepting some of the technical parts) and the entry of Western Technic civilization into a petrified state that, perhaps paradoxically, could continue to progress technically. In this case we would expect a startup to result similar to that seen in the Egyptaic case.

2. This burst of information could have the same effect on Western Technic civilization that the influx of classical Greek knowledge and culture had on early Western civilization: a Renaissance. This response requires that the earth civilization be in a receptive state that is capable of assimilating the influx of knowledge. As Toynbee has pointed out earlier infusions of Greek learning into Europe at the end of the Dark Ages fizzled out because the society of the time was unable to understand and appreciate the knowledge.

Perhaps a better analogy for the impetus given to earth civilization by a burst of knowledge from an extraterrestrial civilization would be the Japanese response to the "opening of Japan" in 1868. The Japanese totally reorganized their government (Meiji Restoration) and society to accommodate a transformation to an advanced industrial economy. The response in this case would be modeled as in the Japanese case with a delta-function force representing the knowledge infusion.

Thus the detection of an extraterrestrial civilization via radio or laser beam signals and the subsequent reception of massive amounts of information in a short time period could have a profound effect on earth civilizations even if there is no "back and forth" communication or physical contact.

Appendix B. Special Relativity Extended to Faster Than Light Motion

This appendix is mathematical in nature. Those not mathematically inclined can skip this appendix. It shows faster than light motion is acceptable. This appendix is an excerpt from our earlier books with some changes in the interests of clarity.

The Theory of Special Relativity is based on the principle that light travels at the same speed in all inertial reference frames. (An inertial reference frame is a reference frame where a particle's speed does not change in the absence of forces.) This principle comes into play when one relates the coordinates of a particle in one inertial coordinate system to the coordinates of a particle in another inertial coordinate system.

The Lorentz group is the essence of Special Relativity. It specifies transformations that relate the coordinates of an event in two coordinate systems that differ by a relative velocity whose magnitude is less than the speed of light.

For example we can imagine an observer in one coordinate system (the "lab" system of an observer) observing an event at time t at the position (x, y, z). Another observer in another coordinate system traveling at a speed v in the x direction relative to the first observer (as depicted in Fig. B.1) observes the same event at time t' at the position (x', y', z'). The relation between the coordinates of the two observers is given by the Lorentz transformation

$$t' = \gamma(t - \beta x/c) \qquad \text{(B.1)}$$
$$x' = \gamma(x - \beta ct)$$
$$y' = y$$
$$z' = z$$

or, in matrix form,

$$X' = \Lambda_L(\omega, \mathbf{u} = (1,0,0))X$$

where $\Lambda_L(\omega, \mathbf{u} = (1,0,0))$ is a matrix representation of the transformation in the x direction symbolized by $\mathbf{u} = (1,0,0)$, where $\beta = v/c$, c is the speed of light, and where $\gamma = (1 - \beta^2)^{-\frac{1}{2}}$.

If β is less than one (sublight speed) then the coordinates of an event are related by eq. B.1. They specify the time and location of an event from the viewpoint of an observer at rest in each coordinate system.

If β is greater than one (superluminal speed) then the coordinates of an event are still related by eq. B.1 but now t' and x' are imaginary numbers. This case is an extension of the Theory of Special Relativity that preserves the principle that the speed of light is the same in all coordinate systems – now including reference frames that are traveling faster than the speed of light relative to a "normal" reference frame.[54]

How can we physically understand coordinates with imaginary values? Well, if we consider the realities of the observer in the "primed" coordinate system it is clear he/she will measure the x' distance with a ruler that measures real numbers, and he/she will measure time t' with a clock that measures real numbers. So the imaginary values of x' and t' in eq. B.1 only appear in the relation between the coordinate systems. If we denote the actual values measured by the primed coordinate system observer as t_r'' and x_r'' then eq. B.1 *for a superluminal relative speed v* becomes

$$t_r'' = i\gamma(t - \beta x/c) \qquad \text{(B.2)}$$
$$x_r'' = i\gamma(x - \beta ct)$$
$$y' = y$$
$$z' = z$$

where $\beta > 1$.

---

[54] A reference frame is another term for inertial coordinate system,

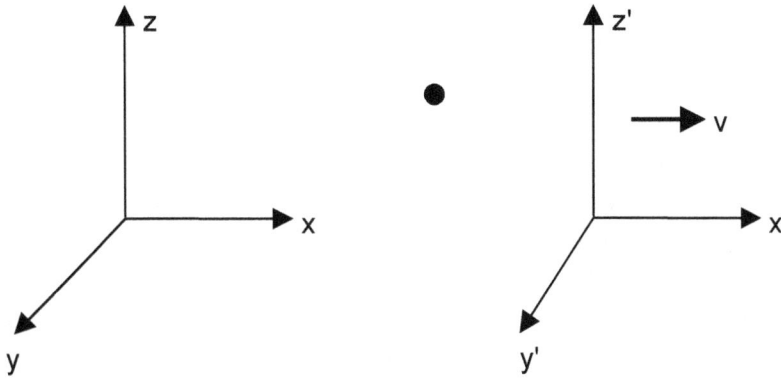

Figure B.1. Two coordinate systems having a relative speed v in the x direction. The black circle represents an event. The "unprimed" coordinate system is the "lab" system. The "primed" coordinate system is the system of an observer moving with speed v along the parallel x and x' axes.

The 4-vector inner product is invariant for $\beta > 1$ for t' and x' as given by eq. B.1:

$$ds^2 = c^2t^2 - x^2 - y^2 - z^2 = c^2t'^2 - x'^2 - y'^2 - z'^2 \qquad (B.3)$$

just as it is for $\beta < 1$. (Note eq. B.3 for double primed coordinates is

$$c^2t^2 - x^2 - y^2 - z^2 = -c^2t''^2 + x''^2 - y'^2 - z'^2 \qquad (B.3a)$$

The physical interpretation of complex coordinates generated by a superluminal transformation is straightforward. Since a superluminal transformation is a linear transformation a straight line whose points are specified by complex numbers will map, in general, into a straight line whose points are specified by real numbers. Thus each point on the lab straight line (a_x, a_y, a_z) is mapped by a superluminal transformation to a three dimensional point whose three coordinate values are complex numbers $(a_{xr}' + ia_{xi}', a_{yr}' + ia_{yi}', a_{zr}' + ia_{zi}')$. An observer in the primed system will, of course, measure real numbers with a ruler between points. These real number coordinate values can be determined by introducing a new transformation that maps the complex numbers generated by superluminal transformations into the real numbers that the primed system observer would measure.

This transformation is of great significance because it leads to a hitherto unstated $SU(2) \otimes U(1) \otimes U(1)$ symmetry. The $SU(2) \otimes U(1)$ part of this new symmetry can be identified as the source of the $SU(2) \otimes U(1)$ symmetry of the ElectroWeak sector of The Standard Model. (See Blaha (2010a).) The additional $U(1)$ symmetry is a source of WIMPs (Weakly Interacting Massive Particles).

We introduce this new transformation by reconsidering the previous simple example (Fig. B.1) wherein a coordinate system is traveling at a speed v in the x direction with respect to the "laboratory" system. The coordinates in the two reference frames are related by eq. B.1. We now define a transformation that maps the real coordinates of the unprimed reference frame to real coordinates in the primed reference frame.

$$\Pi_L(\mathbf{u}) = \begin{bmatrix} -i & 0 & 0 & 0 \\ 0 & -i & 0 & 0 \\ 0 & 0 & 1 & 0 \\ 0 & 0 & 0 & 1 \end{bmatrix} \qquad (B.4)$$

where $\mathbf{u}$ is the unit vector corresponding to the direction of $\mathbf{v}$ (the positive x direction in this example). Using $\Pi_L(\mathbf{u})$ in eq. B.1 we obtain an overall transformation from real coordinates to real coordinates:

$$X'' = \Pi_L(\mathbf{u})X' = \Pi_L(\mathbf{u})\Lambda_L(\omega, \mathbf{u} = (1,0,0)) X$$

or

$$t'' = \gamma_s(t - \beta x)$$
$$x'' = \gamma_s(x - \beta t) \qquad (B.5)$$
$$y'' = y$$
$$z'' = z$$

where $\gamma_s = i\gamma$.

An observer in the primed reference frame would consider his/her time to be real when measured on a clock, and his/her distances along the x-axis to be real when measured with a ruler. Thus eq. B.5 makes good sense physically because in any reference frame observers measure real distances and real times. For this reason we will call transformations of the type of eq. B.5 – from real coordinates to real coordinates – *physical* superluminal transformations.

This simple example generalizes to arbitrary relative velocities **v**. First we note that the Lorentz transformation for a velocity **v** that is a rotation of the velocity in the x-direction (**v** = |**v**|R**u** where R is the relevant rotation matrix) has the form

$$\Lambda_L(\omega, \mathbf{v}) = \mathcal{R}(\mathbf{v}/v, \mathbf{u})\Lambda_L(\omega, \mathbf{u} = (1,0,0))\mathcal{R}^{-1}(\mathbf{v}/v, \mathbf{u}) \tag{B.6}$$

where $\mathcal{R}(\mathbf{v}/v, \mathbf{u})$ is a rotation from the velocity direction **u** to direction **v**/v.

The original transformation (eq. B.5) can be written as

$$\Pi_L(\mathbf{u})\Lambda_L(\omega, \mathbf{u} = (1,0,0)) = \Pi_L(\mathbf{u})\mathcal{R}^{-1}(\mathbf{v}/v, \mathbf{u})\Lambda_L(\omega, \mathbf{v})\mathcal{R}(\mathbf{v}/v, \mathbf{u}) \tag{B.7}$$

Consequently the combined transformation for velocity **v** is

$$\mathcal{R}(\mathbf{v}/v, \mathbf{u})\Pi_L(\mathbf{u})\Lambda_L(\omega, \mathbf{u} = (1,0,0))\mathcal{R}^{-1}(\mathbf{v}/v, \mathbf{u})$$
$$= \mathcal{R}(\mathbf{v}/v, \mathbf{u})\Pi_L(\mathbf{u})\mathcal{R}^{-1}(\mathbf{v}/v, \mathbf{u})\Lambda_L(\omega, \mathbf{v})$$
$$= \Pi_L(\mathbf{v}/v)\Lambda_L(\omega, \mathbf{v}) \tag{B.8}$$

Thus for a Lorentz transformation $\Lambda_L(\omega, \mathbf{v})$ for velocity **v** we see that we can define a subsidiary transformation $\Pi_L(\mathbf{v}/v)$ of the form

$$\Pi_L(\mathbf{v}/v) = \mathcal{R}(\mathbf{v}/v, \mathbf{u})\Pi_L(\mathbf{u})\mathcal{R}^{-1}(\mathbf{v}/v, \mathbf{u}) \tag{B.9}$$

The general form of $\mathcal{R}(\mathbf{v}/v, \mathbf{u})$, is

$$\mathcal{R}(\mathbf{v}/v, \mathbf{u}) = \begin{bmatrix} 1 & 0 & 0 & 0 \\ 0 & & & \\ 0 & & \mathcal{R}_3(\mathbf{v}/v, \mathbf{u}) & \\ 0 & & & \end{bmatrix} \tag{B.10}$$

where $\mathcal{R}_3(\mathbf{v}/v, \mathbf{u})$ is a 3×3 rotation matrix that can be expressed in terms of the generators of the 3-dimensional rotation group as

$$\mathscr{R}_3(\mathbf{v}/v, \mathbf{u}) = \exp(i\boldsymbol{\theta}\cdot\mathbf{J}) \tag{B.11}$$

The rotation angles $\boldsymbol{\theta}$ are real numbers since we are rotating the real vector $\mathbf{u}$ to the real number $\mathbf{v}/v$. Given the form of eq. B.11 then we see that the form of $\Pi_L(\mathbf{v}//v)$ is

$$\Pi_L(\mathbf{v}/v) = \begin{bmatrix} -i & 0 & 0 & 0 \\ 0 & & & \\ 0 & \mathscr{R}_3(\mathbf{v}/v, \mathbf{u})\Pi_{L3}(\mathbf{u})\mathscr{R}_3^{-1}(\mathbf{v}/v, \mathbf{u}) \\ 0 & & & \end{bmatrix} \tag{B.12}$$

where

$$\Pi_{L3}(\mathbf{u}) = \begin{bmatrix} -i & 0 & 0 \\ 0 & 1 & 0 \\ 0 & 0 & 1 \end{bmatrix} \tag{B.13}$$

If we consider the case of an infinitesimal rotation $\boldsymbol{\theta}$ to first order in $\boldsymbol{\theta}$

$$\mathscr{R}_3(\mathbf{v}/v, \mathbf{u}) \simeq I + i\boldsymbol{\theta}\cdot\mathbf{J} \tag{B.14}$$

then

$$\begin{aligned}\Pi_{L3}(\mathbf{v}/v) &= \mathscr{R}_3(\mathbf{v}/v, \mathbf{u}) \, \Pi_{L3}(\mathbf{u}) \, \mathscr{R}_3^{-1}(\mathbf{v}/v, \mathbf{u}) \\ &\simeq \Pi_{L3}(\mathbf{u}) + i\boldsymbol{\theta}\cdot\mathbf{J}\Pi_{L3}(\mathbf{u}) - i\Pi_{L3}(\mathbf{u})\boldsymbol{\theta}\cdot\mathbf{J} \\ &\simeq \Pi_{L3}(\mathbf{u})[I + i\Pi_{L3}^{-1}(\mathbf{u})[\boldsymbol{\theta}\cdot\mathbf{J}, \Pi_{L3}(\mathbf{u})] \end{aligned} \tag{B.15}$$

where $\Pi_{L3}^{-1}(\mathbf{u})$ is the inverse of $\Pi_{L3}(\mathbf{u})$ and [...] represents the commutator. Thus for arbitrary rotations eq. B.15 implies

$$\begin{aligned}\Pi_{L3}(\mathbf{v}/v) &= \mathscr{R}_3(\mathbf{v}/v, \mathbf{u}) \, \Pi_{L3}(\mathbf{u}) \, \mathscr{R}_3^{-1}(\mathbf{v}/v, \mathbf{u}) \\ &= \Pi_{L3}(\mathbf{u}) \exp\{i\Pi_{L3}^{-1}(\mathbf{u}) \, [\boldsymbol{\theta}\cdot\mathbf{J}, \Pi_{L3}(\mathbf{u})]\} \end{aligned} \tag{B.16}$$

We can find the general form of $\Pi_{L3}(\mathbf{v}/v)$ by considering the case of eq. B.4 in more detail. The exponential matrix expression in B.16 can written

$$\begin{aligned}\Pi_{L3}^{-1}(\mathbf{u}) \, [\boldsymbol{\theta}\cdot\mathbf{J}, \Pi_{L3}(\mathbf{u})] &= \Pi_{L3}^{-1}(\mathbf{u}) \, \boldsymbol{\theta}\cdot\mathbf{J} \, \Pi_{L3}(\mathbf{u}) - \boldsymbol{\theta}\cdot\mathbf{J} \\ &= \boldsymbol{\theta}\cdot\mathbf{Q} \end{aligned} \tag{B.17}$$

where

$$Q = \Pi_{L3}^{-1}(u) \, J \, \Pi_{L3}(u) - J = Q' - J \qquad (B.18)$$

The matrices Q_i can be evaluated using eq. B.13 and the matrix representations for the rotation generators J_i: which are equivalent to the SU(2) generators T_i:

$$J_1 = \begin{bmatrix} 0 & 0 & 0 \\ 0 & 0 & -i \\ 0 & i & 0 \end{bmatrix} = T_1 \qquad (B.19)$$

$$J_2 = \begin{bmatrix} 0 & 0 & i \\ 0 & 0 & 0 \\ -i & 0 & 0 \end{bmatrix} = T_2 \qquad (B.20)$$

$$J_3 = \begin{bmatrix} 0 & -i & 0 \\ i & 0 & 0 \\ 0 & 0 & 0 \end{bmatrix} = T_3 \qquad (B.21)$$

The rotation generators satisfy the commutation relations

$$[J_i, J_j] = i\epsilon_{ijk}J_k \qquad (B.22)$$

as do the SU(2) generators:

$$[T_i, T_j] = i\epsilon_{ijk}T_k \qquad (B.23)$$

We can calculate Q' from eqs. B.13 and B.16 – B.20 and obtain

$$Q'_1 = \begin{bmatrix} 0 & 0 & 0 \\ 0 & 0 & -i \\ 0 & i & 0 \end{bmatrix} \qquad (B.24)$$

$$Q'_2 = \begin{bmatrix} 0 & 0 & -1 \\ 0 & 0 & 0 \\ -1 & 0 & 0 \end{bmatrix} \qquad (B.25)$$

$$Q'_3 = \begin{bmatrix} 0 & 1 & 0 \\ 1 & 0 & 0 \\ 0 & 0 & 0 \end{bmatrix} \qquad (B.26)$$

We note that each Q'_i is hermitean and the Q'_i satisfy the commutation relations:

$$[Q'_i, Q'_j] = i\epsilon_{ijk}Q'_k \qquad (B.27)$$

Consequently the set of Q'_i are also equivalent to SU(2) group generators. As a result the exponential factor in eq. B.16:

$$\Pi_{L3}(\mathbf{v}/v) = \Pi_{L3}(\mathbf{u})\exp\{i\boldsymbol{\theta}\cdot(\mathbf{Q}' - \mathbf{J})\} \qquad (B.28)$$

is equivalent to a combination of SU(2) rotations not only in this case but in general for superluminal transformations. The factor $\Pi_{L3}(\mathbf{u})$ is not an SU(2) matrix since its determinant is not +1 but

$$\Pi'_{L3}(\mathbf{u}) = -i\Pi_{L3}(\mathbf{u}) \qquad (B.29)$$

is an SU(2) matrix since

$$\Pi'_{L3}{}^{-1}(\mathbf{u}) = \Pi'_{L3}{}^\dagger(\mathbf{u}) \qquad (B.30)$$
$$\det \Pi'_{L3}(\mathbf{u}) = 1 \qquad (B.31)$$

Thus

$$\Pi'_{L3}(\mathbf{v}/v) = \Pi'_{L3}(\mathbf{u})\exp\{i\boldsymbol{\theta}\cdot(\mathbf{Q}' - \mathbf{J})\} \qquad (B.32)$$

is an SU(2) rotation.

Thus the general form of superluminal transformation from a real set of coordinates to a real set of coordinates is[55]

$$\Pi_L(\mathbf{v}/v)\Lambda_L(\omega, \mathbf{v}) \qquad (B.33)$$

Where

---

[55] The choice of the unit vector $\mathbf{u}$ and the angle vector $\boldsymbol{\theta}$ must be such that applying eq. 2.34 to a real set of coordinates yields a real set of coordinates.

$$\Pi_L(\mathbf{v}/v) = \begin{bmatrix} -i & 0 & 0 & 0 \\ 0 & & & \\ 0 & \Pi_{L3}(\mathbf{u})\exp\{i\,\boldsymbol{\theta}\cdot(\mathbf{Q'}-\mathbf{J})\} & \\ 0 & & & \end{bmatrix} \quad (B.34)$$

by eqs. B.8, B.12 and B.29 – B.32. The Lorentz condition for real to real transformations generalizes to

$$\Lambda(\mathbf{v})^T \Pi_L(\mathbf{v}/v)^\dagger\, G\, \Pi_L(\mathbf{v}/v)\Lambda(\mathbf{v}) = G \quad (B.35)$$

Since superluminal transformations $\Lambda_L(\omega, \mathbf{v})$ transform real coordinates to complex coordinates in general, we can generalize the form of a superluminal transformation to

$$e^{i\phi}\Pi_L(\mathbf{v'}/v')\Lambda_L(\omega, \mathbf{v}) \quad (B.36)$$

where ϕ is a constant phase and $\mathbf{v'}$ is an arbitrary velocity. This generalization will satisfy the generalized Lorentz condition

$$\Lambda(\mathbf{v})^T \Pi_L(\mathbf{v'}/v')^\dagger e^{-i\phi}\, G\, e^{i\phi}\Pi_L(\mathbf{v'}/v')\Lambda(\mathbf{v}) = G \quad (B.37)$$

but the transformation will, in general, yield a complex set of coordinates when applied to a set of real coordinates.

These considerations imply:

1. Any observer in a coordinate system will treat a complex 4-dimensional coordinate system as if it were a real 4-dimensional coordinate system with complex-valued straight lines along each dimension (assuming rectangular coordinates).

2. The transformation $e^{i\phi}\Pi'_{L3}(\mathbf{v}/v)$ is an SU(2)⊗U(1) transformation that takes complex 3-dimensional spatial coordinates to complex 3-dimensional spatial coordinates. In particular straight lines map to straight lines.

3. Physical observations in the observer's coordinate system are invariant under SU(2)⊗U(1) rotations of the spatial coordinates and the multiplication of the time component by an arbitrary phase.

4. The matrix

$$\Pi'_L(\mathbf{v}/v, \chi, \phi) = \begin{bmatrix} e^{i\chi} & 0 & 0 & 0 \\ 0 & & & \\ 0 & e^{i\phi}\Pi'_{L3}(\mathbf{u})\exp\{i\,\boldsymbol{\theta}\cdot(\mathbf{Q}' - \mathbf{J})\} & & \\ 0 & & & \end{bmatrix} \qquad (B.38)$$

(where χ and ϕ are real numbers and $\mathbf{u}$ is a unit vector along any convenient coordinate axis) is an SU(2)⊗U(1)⊗U(1) transformation that transforms complex 4-dimensional coordinates to complex 4-dimensional coordinates. Note, $\Pi_L(\mathbf{v}/v) = \Pi'_L(\mathbf{v}/v, 3\pi/2, \pi/2)$ is a special case of $\Pi'_L(\mathbf{v}/v, \chi, \phi)$. Due to the manifest form of eq. B.38 we see

$$\Pi'^{\mu}_{L\,\alpha}{}^*\Pi'^{\mu}_{L\,\beta} = [\Pi'_L{}^\dagger\Pi'_L]_{\alpha\beta} = I_{\alpha\beta} \qquad (B.39)$$

(with an implied sum over μ) or, in matrix form,

$$\Pi'_L{}^\dagger \Pi'_L = I \qquad (B.40)$$

and also[56]

$$\Pi'_L{}^\dagger G\Pi'_L = G \qquad (B.41)$$

5. Complex coordinate values of the type generated by superluminal transformations are transformable to real coordinates by a transformation of the form of eq. B.38 – an SU(2)⊗U(1)⊗U(1) transformation. The complex coordinates are thus physically equivalent to corresponding real coordinate values in the sense that an observer in that frame would automatically use the real coordinates so obtained

---

[56] Eq. B.35 is close to the defining condition for a Lorentz group element but the presence of complex conjugation rather than a transpose means Π'_L is outside the real and complex Lorentz groups.

since rulers and clocks always measure real spatial coordinates and times.

6. The complex coordinates of any point obtained through a superluminal transformation can be transformed to a real set of coordinates by a transformation of the form of eq. B.38.

$SU(2) \otimes U(1) \otimes U(1)$ invariance can be shown to lead to the $SU(2) \otimes U(1) \otimes U(1)$ ElectroWeak sector of the Standard Model with WIMPs which can be restricted to an $SU(2) \otimes U(1)$). See Blaha (2010a).

7. The form of eq. B.37 implies that the space-time group of the Standard Model is

$$SU(2) \otimes U(1) \otimes U(1) \otimes L_c \qquad \text{(B.37a)}$$

where L_c is the complex Lorentz group. This group is a subgroup of the complex group GL(4). Space-time, in general has 4 complex dimensions. These dimensions can be treated as real dimensions because of the equivalence of complex coordinates to real coordinates under $SU(2) \otimes U(1) \otimes U(1)$ transformations.

8. The $SU(2) \otimes U(1) \otimes U(1)$ symmetry is extended in later works such as Blaha (2011d) to $SU(3) \otimes SU(2) \otimes U(1) \otimes SU(2) \otimes U(1)$. The factors in this product of groups all appear in U(4) (also shown in Blaha (2011d)). Since a U(4) transformation exists that will transform any complex 4-vector into a real-valued 4-vector, the product can also transform any complex 4-vector into a real-valued 4-vector. In general, the transformation will be position dependent. Consequently, the relevant group that reduces complex space-time to real space time is the *local* group $SU(3) \otimes SU(2) \otimes U(1) \otimes SU(2) \otimes U(1)$. Its transformations are dependent on space-time positions and vary as the position changes. In this regard they are like the $SU(3) \otimes SU(2) \otimes U(1) \otimes SU(2) \otimes U(1)$ local group appearing in our extension of The Standard Model.

Appendix C. Time and Space Contraction and Dilation

In ordinary Lorentz transformations a moving ruler will appear to be shorter in the direction of its motion when measured in another reference frame. This phenomenon is called *Lorentz contraction*. In ordinary Lorentz transformations time intervals will appear to be longer when measured in another reference frame. This phenomenon is called *time dilation.*

In superluminal (faster than light) transformations contraction and dilation are more complicated as we show in this section.

C.1 Superluminal Length Dilation/Contraction

In the case of a superluminal transformation we find *superluminal length contraction or dilation* can occur depending on the relative velocity. Consider the case of the transformation of eq. B.1 above, which relates the primed reference frame traveling at speed v in the positive x direction to the unprimed reference frame. A ruler perpendicular to the x-axis will have the same length in both reference frames if its endpoints are simultaneously measured – perhaps by photographing it. The y and z equations in eqs. B.1 specify this fact.

If the ruler is at rest in the primed reference frame and parallel to the x' axis, then a simultaneous measurement of its endpoints at the same time t_0 by an observer in the unprimed reference frame (perhaps by photographing it) will reveal either *length contraction and dilation* depending on the value of β. If the length is $L' = x'_2 - x'_1$ in the primed frame and $L = x_2 - x_1$ in the unprimed frame, then the equations:

$$x'_1 = \gamma_s(x_1 - \beta ct_0) \tag{C.2}$$
$$x'_2 = \gamma_s(x_2 - \beta ct_0) \tag{C.3}$$

where $\gamma_s = i\gamma$ imply

$$L' = \gamma_s L = (\beta^2 - 1)^{-\frac{1}{2}} L \qquad \text{(C.4)}$$

Thus we have three cases:

Case 1: $\beta \in <1, \sqrt{2}>$: $\qquad\qquad$ $L < L'$ $\quad$ Contraction $\qquad\qquad$ (C.5)

Case 2: $\beta = \sqrt{2}$: $\qquad\qquad\qquad$ $L = L'$ $\quad$ Equality $\qquad\qquad\quad$ (C.6)

Case 3: $\beta \in <\sqrt{2}, \infty>$: $\qquad\qquad$ $L > L'$ $\quad$ Dilation $\qquad\qquad\quad$ (C.7)

Thus $\beta = v/c = \sqrt{2}$ marks the point of change from a contraction of lengths[57] to a dilation of lengths. The dilation feature of superluminal motion, first noted in Blaha (2007a), has no counterpart in sublight motion.

$\qquad$ *The effect of the change at $\beta = \sqrt{2}$ on a starship is startling.* Imagine the primed coordinate system is that of the starship and the unprimed coordinate system is that of the earth. (The motion of the earth is small and can be neglected relative to the high speed of the starship.) If the starship velocity moving on a straight line away from the earth is such that β is between 1 and $\sqrt{2}$ then a yardstick on the starship will appear to be shorter in the *earth* coordinate system – Lorentz contraction. If the starship velocity is such that β is greater than $\sqrt{2}$ then a yardstick on the *starship* will appear to be longer in the earth coordinate system – dilation. More interestingly, if the starship travels 1 light year in its coordinate system it will actually have traveled more than 1 light year in the earth's coordinate system – possibly *much* more than 1 light year in the earth's coordinate system if the starship's speed has a β that is much greater than $\sqrt{2}$. Thus, for example, a 5 light year distance measured on earth will be *less* than the same distance measured on a starship, namely 2.89 light years.

C.2 Superluminal Time Contraction/Dilation

$\qquad$ In the case of a superluminal transformation *superluminal time contraction* is a possibility.[58] Consider again the case of the

---

[57] This is the case for sublight Lorentz transformations.
[58] Blaha (2007a).

transformation of eq. B.1 relating the primed reference frame traveling at speed v in the positive x direction to the unprimed reference frame. Consider the time interval between two events occurring at the same point x'_0 in the primed reference frame. From the viewpoint of an observer in the unprimed frame the events take place at different points x_1 and x_2. If the time interval is $T' = t'_2 - t'_1$ in the primed frame and $T = t_2 - t_1$ in the unprimed frame, then the inverse transformation to eq. B.1 gives:

$$t_1 = \gamma_s(t'_1 + \beta x'_0/c) \tag{C.8}$$
$$t_2 = \gamma_s(t'_2 + \beta x'_0/c) \tag{C.9}$$

and implies

$$T = \gamma_s T' = (\beta^2 - 1)^{-\frac{1}{2}} T' \tag{C.10}$$

Again there are three cases:

Case 1: $\beta \in \langle 1, \sqrt{2} \rangle$: $T > T'$ Dilation   (C.11)

Case 2: $\beta = \sqrt{2}$: $T = T'$ Equality   (C.12)

Case 3: $\beta \in \langle \sqrt{2}, \infty \rangle$: $T < T'$ Contraction   (C.13)

The time interval in the unprimed (earth) frame can be less than, equal to, or greater than the time interval in the primed frame when the events take place at the same spatial point.

Thus superluminal transformations are more complex than sublight Lorentz transformations (which only have time dilation) with respect to time dilation and contraction.

The effect of the time change at $\beta = \sqrt{2}$ on a starship is also startling. Again imagine the primed coordinate system is that of the starship and the unprimed coordinate system is that of the earth. If the starship velocity moving on a straight line away from the earth is such that β is between 1 and $\sqrt{2}$ then a time interval on the starship will appear to be longer in the earth coordinate system – time dilation. Or, stating it otherwise, time intervals on the starship will appear to be shorter than they appear on earth. A person on a starship would thus age more slowly from the point of view of a person on earth.

But if the starship velocity is such that β is greater than $\sqrt{2}$ then a time interval on the starship will appear to be shorter in the earth coordinate system – time contraction. If the starship travels for 1 year in its coordinate system it will actually have traveled less than a year in the earth's coordinate system – possibly much less than a year in the earth's coordinate system – if the starship's speed has a β that is much greater than $\sqrt{2}$. *Thus people on the starship would appear to age more quickly than on earth. At high speed a starship crew could age very rapidly. Thus the need for suspended animation.*

Appendix D. Faster Than Light Starship Dynamics

This appendix is mathematical in nature. Those not mathematically inclined can skip this appendix. The non-mathematical reader need only know that faster than light starships are possible.

D.1 Exceeding the Speed of Light

In standard sublight relativistic dynamics the speed of a massive object cannot exceed the speed of light if a force applied to the object is real. In this section we will consider the case of a *complex-valued* force applied to an object (complex thrust) that causes the object to attain a complex velocity whose real part can exceed the speed of light. Complex valued forces (tachyonic forces) have not been experimentally found in Nature as yet. However the motion of a particle inside a Black Hole is tachyonic. However we have seen solid evidence of tachyons in chapter 6. In Blaha (2010a) we showed that neutrinos and d type quarks are tachyons. Up-type quarks also have complex 3-momenta. The real part of their speed can be accelerated to faster than light or have speeds below the speed of light. If we harness quarks to create rocket thrust then we can have a mechanism for complex-valued thrust that can power starships faster than the speed of light.

Since a complex-valued rocket thrust will generate a complex-valued velocity, and movement in space, the physical interpretation of complex velocities and distances must be addressed. In appendix B we showed that a superluminal transformation maps points with real coordinate values in one coordinate system to points with complex coordinate values in the target coordinate system. We then showed that the complex-valued coordinate points in the target coordinate system could be "rotated" to real valued coordinates using $\Pi_L(\mathbf{v}/v)$ (eq. B.34). Thus the combined superluminal transformation and $\Pi_L(\mathbf{v}/v)$

transformation $\Pi_L(\mathbf{v}/v)\Lambda(\mathbf{v}/v)$ maps real coordinates to real coordinates. Complex coordinates are then merely an artifact of superluminal transformations.

However when we consider the path of a rocket with complex thrust that starts from a spatial point with real coordinates and, as it accelerates, traverses complex-valued spatial points a new issue arises: What is the physical meaning of these complex-valued spatial coordinates. Unlike the previous case of superluminal transformations one cannot simply use a global transformation to change the complex-valued points to points with real coordinate values. This is particularly clear if one considers a three point configuration: the earth, the starship and the destination star.

The earth and star have real valued coordinates in the earth coordinate system. The starship in transit has complex coordinates at each point of its journey in starship coordinates. In general, there is no global transformation that will make the coordinates of all three points real-valued.

Therefore we conclude that complex coordinates are physically meaningful in this type of situation where one "point" is moving with a complex velocity. On this basis we will assume that space is three-dimensional with complex coordinate values in general. *But* using the local transformations of the Reality group described briefly in point 8 of appendix B, namely local $SU(3){\otimes}SU(2){\otimes}U(1){\otimes}SU(2){\otimes}U(1)$, these transformations map complex 4-dimensional space to real-valued space – exactly the kind of space-time that we see.

Why haven't the complex values of coordinates been noticed before? Because objects with complex velocities have not been created and/or seen by Man. To give an object a complex velocity we need either a highly curved space-time region (such as a Black Hole) with an event horizon that encloses the object so that we can't see it; or to accelerate an object with a complex-valued force or thrust giving it a complex velocity and consequently a trajectory in complex coordinates.

This second possibility can only be achieved with tachyon thrust or force generated by accelerating quarks and gluons. Neutrinos, although tachyonic, only interact via the Weak interaction and have strictly real momenta. Thus they are not capable of generating a complex thrust.

Quarks are confined within protons and neutrons. To create macroscopic regions containing quarks we need collisions at enormous energies. We are just entering the experimental stage where this possibility can be realized. RHIC at Brookhaven National Laboratory and LHC at CERN have started creating quark-gluon fluids by colliding heavy ions such as gold and lead ions. Evidence for tachyonic quarks within the collision regions will hopefully be forthcoming soon.

Then a superluminal, tachyon drive starship with complex thrust is possible.

D.2 Superluminal Starship Dynamics

In this section we will consider a constant, propulsive force in a starship's rest frame that drives the starship from a sublight velocity to a superluminal velocity. The key factor in achieving a superluminal speed is evading the singularity in γ at $v/c = 1$. We accomplish this goal by having a complex force – a force with a real and imaginary part – that generates a complex acceleration, and thus a complex velocity, that "goes around" the singularity in γ in the complex velocity plane. We assume that an "instantaneous" Lorentz transformation relates the earth reference frame and the starship reference frame.

We assume the starship's thrust is in the direction of the positive x' (and x) axis. We also assume for simplicity that the mass of the starship is constant. (The starship engine uses a small amount of fuel relative to the starship's total mass.) The starship (primed coordinates) and earth (unprimed coordinates) coordinates have parallel axes as in Fig. B.1. We assume the spatial force is in the positive x direction

$$\mathbf{F'} = g\hat{\mathbf{x}} \qquad\qquad (D.1)$$

where g is assumed to be a complex constant.

The fourth component of the force (since force is a Lorentz 4-vector) is zero in the rocket's rest frame:

$$F'^0 = 0 \qquad\qquad (D.2)$$

Applying the inverse of the Lorentz transformation eq. B.1 we find the force in the earth rest frame is

$$F^0 = \gamma(F'^0 + \beta F'^x/c) = \gamma\beta F'^x/c = \gamma v g/c^2 \qquad (D.3)$$
$$F^x = \gamma(F'^x + \beta c F'^0) = \gamma F'^x = \gamma g$$
$$F^y = F^z = 0$$

where $\beta = v/c$, c is the speed of light, and $\gamma = (1 - \beta^2)^{-\frac{1}{2}}$ as before. We again use the superscripts x, y, and z to identify the components of the spatial force. The spatial momentum of an object of mass m is

$$\mathbf{p} = \gamma m \mathbf{v} \qquad (D.4)$$

and the dynamical equation of motion in the earth's rest frame is

$$d\mathbf{p}/dt = \mathbf{F} \qquad (D.5)$$

resulting in

$$dp^x/dt = \gamma g \qquad (D.6)$$

with

$$dp^y/dt = dp^z/dt = 0 \qquad (D.7)$$

The differential equation resulting from eq. D.5 is

$$d(\gamma v)/dt = \gamma g/m \qquad (D.8)$$

assuming, as stated earlier, the fuel used is small[59] compared to the starship's mass. The solution of eq. D.8 is

$$v = c - 2c/(1 + ((c + v_0)/(c - v_0))\exp[2g(t - t_0)/(mc)]) \qquad (D.9)$$

where the velocity is v_0 at time t_0. If we take account of the decreasing fuel mass then we expect the velocity will increase somewhat more rapidly than eq. D.9. *Since v is a complex number, due to the complex*

---

[59] Perhaps of the order of 10% – 20% of the starship mass.

acceleration, the singularity at v = c in γ is avoided and the starship can surpass the speed of light with no difficulty.

Integrating eq. D.9 we find the x distance traveled is

$$x = x_0 - c(t - t_0) + (mc^2/g)\ln((1 - v_0/c + (1 + v_0/c)\exp[2g(t - t_0)/(mc)])/2)$$
$$(D.10)$$

The complexity of g causes v and x to be complex. The complexity of x raises the question of its interpretation. The Reality group introduced in Blaha (2011d) and (2012a) furnishes the required interpretation. Since no spatial rotation takes place in this situation, and we are still using the earth coordinates, the relevant Reality group transformation only transforms the complex x value to a real value equal to its absolute value.[60] The matrix form of the transformation is

$$\Pi_L = \begin{bmatrix} 1 & 0 & 0 & 0 \\ 0 & e^{i\varnothing} & 0 & 0 \\ 0 & 0 & 1 & 0 \\ 0 & 0 & 0 & 1 \end{bmatrix} \qquad (D.11)$$

where $\varnothing$ is the phase of x. The real-valued coordinate x_p is the value of the distance traveled from earth which, of necessity, must be a real number measurable by a very long yardstick in principle.

$$x_p = |x| = xe^{-i\varnothing} \qquad (D.10a)$$

As a result superluminal travel to a distant star (or galaxy eventually) requires three phases in general. In the first phase (phase I) the starship accelerates with a value for the thrust g that enables it to reach a high complex velocity v_h whose real part is much greater than

---

[60] The Reality group appears in two roles. If a superluminal transformation is made between coordinate systems then a Reality group transformation is needed to make the target superluminal system coordinates into physical, real-valued coordinates such as an observer in that coordinate system would measure. Secondly if a dynamical state evolves in such a way that some of the coordinates become complex then a Reality group transformation is used to change complex coordinate values to physical, real-valued coordinates. Appendix D contains examples.

the speed of light (eq. D.9). In the second phase (phase II) the starship coasts with speed v_h at a constant high speed to a point "not far" from the destination. In the third phase (phase III) the starship engines are turned on and the starship decelerates to a low speed at its destination solar system. See Fig. D.1.

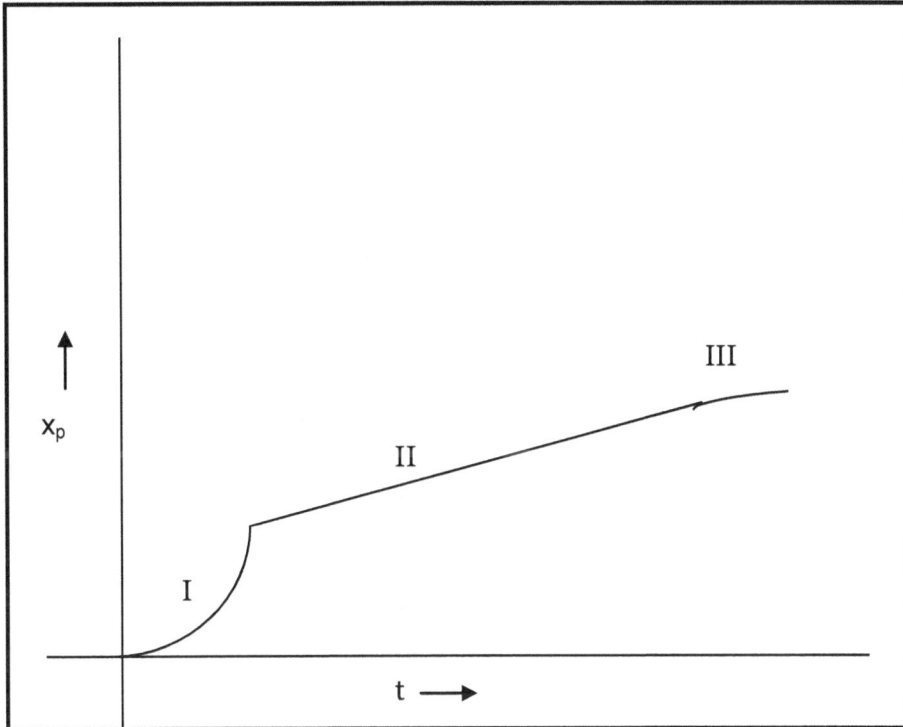

Figure D.1. Rough depiction of the travel of a starship in the x dimension. The starship starts out in real space at x = 0. It accelerates in phase I. After reaching cruising speed it turns off its superluminal engines (phase II), and coasts until it is near its destination. When near its destination it turns the superluminal engines back on (phase III), decelerating it to low speed.

D.3 Achieving High Superluminal Starship Velocities

To achieve *much* faster than light motion the constant force value g required must satisfy a special set of conditions. These conditions emerge from a consideration of the denominator of eq. D.9:

$$1 + ((c + v_0)/(c - v_0))\exp[2g(t - t_0)/(mc)] \qquad (D.12)$$

If $v_0 < c$ then the denominator can only be "infinite" if g is a complex number. If $v_0 > c$ then the denominator can be "infinite" if g is a real or a complex number. The following cases are of interest:

$\underline{c < v_0}$

In this case g is real and the denominator zero is specified by

$$2g(t - t_0)/(mc) = \ln((v_0 - c)/(c + v_0)) < 0$$

implying g is negative and real since $t - t_0 > 0$. A negative g value corresponds to deceleration of the starship if it is traveling faster than light.

$\underline{c > v_0}$

In this case the denominator zero satisfies

$$2g(t - t_0)/(mc) = \ln((v_0 - c)/(c + v_0)) = \text{a complex number}$$

implying g must be complex. This case corresponds to a starship accelerating from a small speed to light speed.

We will consider the general case of complex g since this type of thrust allows us to go from very low speed to much faster than light speed – the general goal of starship travel. Let

$$g = g_1 + ig_2 \qquad (D.13)$$

If we wish the velocity to get very large (approach infinity) after some acceleration time interval $\triangle t = t_1 - t_0$ we set

$$1 + ((c + v_0)/(c - v_0))\exp[2g\triangle t/(mc)] = 0 \qquad (D.14)$$

with the result

$$g_2 = (mc/(2\triangle t))\{n\pi + \text{Im} \ln[(c - v_0)/(c + v_0)]\} > 0 \qquad (D.15)$$

and

$$g_1 = (mc/(2\triangle t))\ \text{Re}\ \ln[(c - v_0)/(c + v_0)] < 0 \qquad (D.16)$$

where n is an odd, positive integer, since v_0 is complex in general. Eqs. D.15 and D.16 enable the real and imaginary parts of the velocity (and thus the absolute value of the velocity) to become infinite as the time interval approaches $\triangle t$. We assume n = 1 in the following discussions. Substituting in eq. D.9 we obtain

$$v = c\{1 - 2/[1 + ((c + v_0)/(c - v_0))^{1 - (t - t_0)/\triangle t}\ e^{in\pi(t - t_0)/\triangle t}]\} \qquad (D.17)$$

We will now approximate eq. D.9's denominator as it approaches zero. Letting $t = t_1 + \tau$ where τ is small, and letting $\triangle t = t_1 - t_0$ then eq. D.9 becomes

$$\begin{aligned}
v &= c\{1 - 2/(1 + ((c + v_0)/(c - v_0))\exp[2g(\triangle t + \tau)/(mc)])\} \\
&= c\{1 - 2/(1 - \exp[2g\tau/(mc)])\} \\
&\simeq c\{1 - 2/(1 - (1 + 2g\tau/(mc))\} \\
&\simeq c\{1 + (mc/g)(1/\tau)\} \\
&\simeq (g^*mc^2/|g|^2)(1/\tau) \qquad (D.18)
\end{aligned}$$

Eq. D.18 shows

- For small negative τ both the real and imaginary parts of v approach $+\infty$ as $\tau \to 0$ from below.
- For small positive τ both the real and imaginary parts of v approach $-\infty$ as $\tau \to 0$ from above.

as displayed in Figs. D.2, D.3 and D.4.

The singular behavior as $\tau \to 0$ from above or below requires some explanation. It is not like the singular behavior as $v \to c$ seen in Special Relativity. Rather, it is a result of the use of the time coordinate of the earth's coordinate system. For if we transform earth time t to starship time t_r'' using eq. B.2

$$t_r'' = i\gamma(t - \beta x/c) \qquad (B.2)$$

and determine the time contraction of an interval T from eq. C.10 then from

$$T = \gamma_s T'' = (\beta^2 - 1)^{-\frac{1}{2}} T'' \qquad \text{(C.10)}$$

we see the starship time interval is

$$T'' \approx vT/c \qquad \text{(D.19)}$$

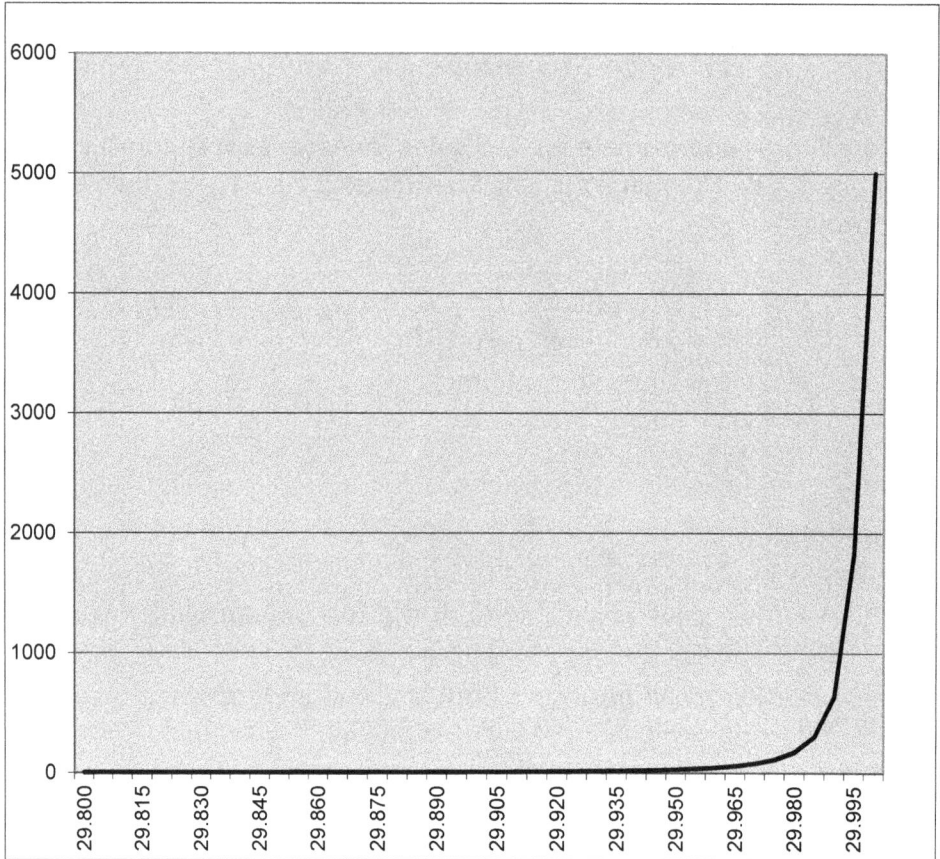

Figure D.2. Sample plot of the real part of the velocity of a starship on its 29th and 30th earth day of travel up to 5,000c. Time is measured in earth coordinates. Velocity is measured in units of c, the speed of light.

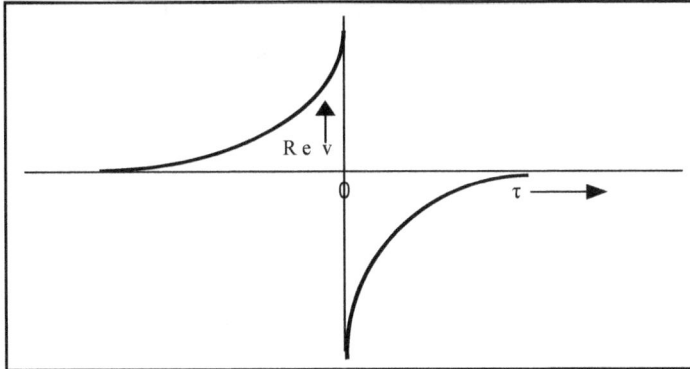

Figure D.3. Sample qualitative plot of Re v near the singularity at τ = 0.

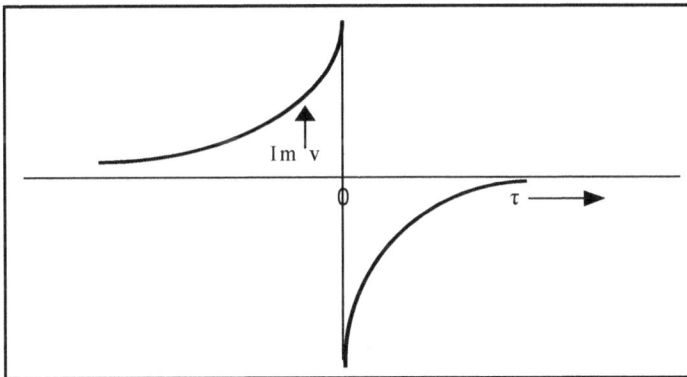

Figure D.4. Sample qualitative plot of Im v near the singularity at τ = 0.

Thus as v approaches infinity the starship time interval T'' grows to infinity as well. A starship will never reach the singular point τ = 0 in a finite time. But, depending on fuel availability, it can reach speeds much, much faster than the speed of light.

   The acceleration of the starship in the starship's coordinate system is

$$a' = F'^x/m = g/m \qquad (D.20)$$

while the acceleration, a, in the earth's coordinate system is given by the derivative of eq. D.9.

$$a = dv/dt$$
$$= 4(g/m)((c + v_0)/(c - v_0))\exp[2g(t - t_0)/(mc)]/\{1 +$$
$$+ ((c + v_0)/(c - v_0))\exp[2g(t - t_0)/(mc)]\}^2 \qquad (D.21)$$

At $t = t_1 + \tau$ we see

$$a \simeq -mc^2/(g\tau^2)[1 + 2g\tau/mc] \approx -mc^2/(g\tau^2) \qquad (D.22)$$

in the earth's reference frame.[61] Thus the starship appears to have its acceleration approach infinity as τ approaches zero in the earth's coordinate system. This apparent problem does not exist in the starship reference frame where the acceleration is constant.

The crew can experience accelerations up to four times earth's normal gravitational acceleration without harm. If they are in suspended animation they should be able to withstand higher accelerations – perhaps eight times earth's gravitational acceleration.

The fuel expended as the earth time interval τ approaches zero in this case must approach infinity since there is a constant acceleration for an infinite time in the starship coordinate system. The acceleration is generated by the propellant exhaust. In this case we assume the fuel expended to accelerate to a high velocity is small compare to the starship's total mass.

We will examine the case where the fuel is not a small fraction of the starship mass later in this appendix.

D.4 Constant Superluminal Starship Travel

Assuming the starship has accelerated to an enormous *real* speed such as a speed between 5000c and 30,000c we can turn off the superluminal engines. The starship then moves at this constant speed in the absence of forces gravity, retarding effects of space dust, and other forces).

---

[61] Approximation eq. D.22 is the derivative of eq. D.18 as expected.

Consider a starship speed of 5000c. Any place in the galaxy is a short travel time away. And nearby galaxies are reachable as well. Figure D.7 shows the time required to reach various interesting destinations at a much higher speed of 30,000c.

Destination	Distance (ly)	Approximate Travel Time (years)
To the other end of the Milky Way Galaxy	100,000	3
To the Center of the Milky Way	30,000	1
Large Magellenic Galaxy	150,000	5
Small Magellenic Galaxy	200,000	7
Andromeda Galaxy	2,000,000	70

Figure D.5. "Coasting" part of travel time to various destinations at a real velocity of 30,000c.

Since much, much higher "coasting" velocities are also possible almost the entire visible universe becomes accessible to Mankind if we can boost quark-gluon exhaust velocities to very large values. Mankind then has an incredible future if it has the will to seize it.

D.5 Deceleration of a Tachyonic Starship to Sublight Speeds

Eventually all journeys end so we will now examine the deceleration of a starship as it approaches its destination. We turn on the superluminal engine. The thrust is reversed (g → –g) to decelerate as the target star system is approached.

D.6 Fuel Consumption

The acceleration of a rocket of mass m with a propellant exhaust speed v_e in the rocket's rest frame is given by

$$dv'/dt' = (v_e/m) \, dm/dt' \qquad (D.23)$$

and thus the constant g of eq. D.1 is

$$g = mdv'/dt' = v_e \, dm/dt' \qquad (D.24)$$

Since we intend to generate the thrust with a quark-gluon plasma producing an extremely high-energy exhaust we will *choose* the value of the starship acceleration to be equal to the acceleration due to gravity at the earth's surface times $8(1 + i)$:[62]

$$g/m = 8(1 + i)g_E = 8(1 + i)980 \text{ cm/sec}^2 \qquad (D.25)$$

where m is the mass of the starship. If we specify an exhaust velocity v_e

$$v_e = -1000(c + ic) \qquad (D.26)$$

which is a reasonable choice for the exit speed thrust of the fireball then

$$dm/dt' = -2.61 \times 10^{-10} \text{ m} \qquad (D.27)$$

If the starship weighs 10,000 metric tons[63] then

$$dm/dt' = -2.61 \text{ gm/sec} \qquad (D.28)$$

From the viewpoint of rockets, dm/dt' is a small quantity. But due to time dilation the cumulative effect of dm/dt' in multi-year travel in starship time is a relatively large amount of fuel.

However if the starship can use processed material from asteroids and moons to make fuel then the limitation on travel imposed by fuel consumption can be circumvented. The fuel need not be composed of specific elements such as lead or uranium but could be spherules composed of a variety of materials if the starship engine were designed to handle such a variety of spherules.

The amount of fuel used per unit time would appear to be acceptable for quark-gluon plasma production for an ion drive. Currently minuscule amounts of plasma are created with ion-ion collisions. Colliding spherules of 1 milligram mass would require a not unreasonable collision rate of 1305 nominal collisions per second.

---

[62] Eight g's in astronaut terminology.
[63] About one-fifth the mass of the ship Queen Elizabeth.

Appendix E. Acceleration and Confinement of Superluminal Particles with Electromagnetic Interactions

This appendix is mathematical in nature. Those not mathematically inclined can skip this appendix. The non-mathematical reader need only know that superluminal particles including those in a quark-gluon plasma can be accelerated and manipulated by electromagnetic fields.

In this appendix we will consider the acceleration, and the bending, of superluminal particles by electromagnetic fields.[64] Since superluminal particles such as quarks have complex-valued velocities they have significantly different behavior under electromagnetic forces. This appendix is written with a view to the future when much more powerful magnets and electric fields can effectively manipulate quark-gluon fireballs. Quarks, having charge, can be accelerated by electric and magnetic fields. They can drag (zero electric charge) gluon fields along with them.

E.1 Lorentz Force Equations

The key equation for the motion of a charged particle under the influence of an electromagnetic field is the Lorentz force equations:

---

[64] The reader may feel that many of the parts of this appendix are somewhat elementary but upon examination will se that the superluminal nature of the particles introduces subtle points of difference from the corresponding subluminal kinematics. **We note that the discussions of superluminal particles apply to quasi-free quarks within a quark gluon plasma or fluid such as is generated in high energy hadron collisions at RHIC or the LHC at present. The quarks are confined to a region that ranges from spherical to elongated. This region has complex spatial coordinates and momenta. The region could be viewed as a "bag" – that is a bubble-like region with a surface. The quark acceleration topics in this appendix implicitly take place within such a conceptual bag.**

$$d\mathbf{p}/dt = qe(\mathbf{E} + \mathbf{v} \times \mathbf{B}/c) \qquad \text{(E.1)}$$
$$dE/dt = qe\mathbf{v}{\cdot}\mathbf{E} \qquad \text{(E.2)}$$

where $\mathbf{E}$ and $\mathbf{B}$ are electric and magnetic fields, and qe is the charge of the particle.

An issue that arises is the requirement that the energy of a particle be real. If the particle velocity is complex as it is in general for quarks:[65]

$$\mathbf{v} = \mathbf{v}_r + i\mathbf{v}_i \qquad \text{(E.3)}$$

where

$$\mathbf{v}_r{\cdot}\mathbf{v}_i = 0 \qquad \text{(E.4)}$$

then for the particle energy to be *real* as time evolves

$$\mathbf{v}_i{\cdot}\mathbf{E} = 0 \qquad \text{(E.5)}$$

must hold by eq. E.2 since the electric field $\mathbf{E}$ is real. If $\mathbf{v}_i{\cdot}\mathbf{E} \neq 0$ then the energy becomes complex as time progresses. Then the interpretation of a complex energy becomes an issue. The issue is resolved by noting that a superluminal transformation at time t to a coordinate system moving at a velocity of $\mathbf{v}_i(t)$ makes the imaginary velocity in the new coordinate system zero so that dE/dt is instantaneously real. Application of the Reality group transformation $\Pi'_L = \text{diag}(e^{i\chi}, 1, 1, 1)$ (See eq. B.38) to the transformed particle 4-momentum with a suitable choice of χ will make the energy E(t) a real number. Thus the physical interpretation of a complex particle energy is clear when the coordinate system is boosted to a coordinate system by a superluminal transformation followed by an appropriate choice of Π'_L.

E.2 Superluminal Particle Acceleration under a Constant Electric Force

We will consider the case of a superluminal particle moving in a constant electric field where the real part of the particle momentum is parallel to the electric field (and thus the imaginary part of the velocity

---

[65] Blaha (2010a) and earlier works.

is perpendicular to the electric field). Since the electric force is real, the Lorentz force equation eq. E.1 becomes

$$d\mathbf{p}_r/dt = qe\mathbf{E} \tag{E.6}$$
$$d\mathbf{p}_i/dt = 0$$

where $\mathbf{p}_r$ is the real part and $\mathbf{p}_i$ the imaginary part, of the 3-momentum. Eq. E.6 implies the imaginary part of the particle 3-momentum is constant in time. However the imaginary velocity changes in magnitude and direction with time. [66] The real part

$$\mathbf{p}_r = \gamma m \mathbf{v}_r \tag{E.7}$$

when inserted in eq. E.6 gives the dynamical equation

$$md(\gamma\mathbf{v}_r)/dt = qe\mathbf{E} \tag{E.8}$$

where

$$\gamma = (1 - \beta^2)^{-\frac{1}{2}} \tag{E.9}$$

with

$$\beta = |\mathbf{v}|/c = (\mathbf{v}\cdot\mathbf{v})^{\frac{1}{2}}/c = (\mathbf{v}_r\cdot\mathbf{v}_r - \mathbf{v}_i\cdot\mathbf{v}_i)^{\frac{1}{2}}/c \tag{E.10}$$

Eq. E.8 integrates to

$$\mathbf{p}_r(t) = \mathbf{p}_{r0} + qe\mathbf{E}(t - t_0) = \gamma m \mathbf{v}_r(t) \tag{E.11a}$$

where $\mathbf{p}_{r0}$ is the momentum at time t_0. Note the imaginary part is constant and equal to its initial value in this case

$$\mathbf{p}_i(t) = \mathbf{p}_{i0} = \gamma m \mathbf{v}_i(t) \tag{E.11b}$$

The change in energy with time can be obtained from eq. E.2:

$$E(t) = E_0 + qe\mathbf{E}\cdot(\mathbf{r}(t) - \mathbf{r}_0) \tag{E.12}$$

---

[66] Note γ depends on time due to its dependence on the instantaneous velocity.

where E_0 is the energy at time t_0, $r(t)$ is the position at time t, and r_0 is the position at time t_0:

$$r(t) = \int_{t_0}^{t} dt\ v(t) \qquad (E.13)$$

Note $r(t)$ is a complex 3-vector in general since the velocity is a complex 3-vector. Eq. E.12 is identical in form to the energy calculated for "normal" particles, the change in particle energy equals the change in potential energy, but differs in that $E(t)$ is complex in the present case since the coordinate difference $r(t) - r_0$ is complex in general.

We conclude this subsection by noting again that the imaginary part of a particle's spatial momentum is also changed by an electric field in general. We can calculate the real and imaginary velocities from eqs. E.11. Defining

$$\alpha(t) = \gamma |v_r(t)| = |p_{r0} + qeE(t - t_0)|/m \qquad (E.14)$$

$$\delta = \gamma |v_i(t)| = |p_{i0}|/m \qquad (E.15)$$

we find after some algebra

$$|v_r(t)| = \alpha/[1 + (\alpha^2 - \delta^2)/c^2]^{\frac{1}{2}} \qquad (E.16)$$

$$|v_i(t)| = \delta/[1 + (\alpha^2 - \delta^2)/c^2]^{\frac{1}{2}} \qquad (E.17)$$

for $c^2 + (\alpha^2 - \delta^2) > 0$. Although the imaginary momentum is constant, the imaginary velocity changes magnitude, but not its direction perpendicular to E in this case, due to its appearance in γ. For sublight motion, as $t \to \infty$, $\alpha(t) \to \infty$ and $|v_i(t)| \to 0$ while $|v_r(t)| \to c$ in conformity with the Special Theory of Relativity limit on particle speeds to less than the speed of light.

However if $c^2 + (\alpha^2 - \delta^2) < 0$ then as t approaches a value such that $c^2 + (\alpha^2 - \delta^2) \to 0$, then $|v_r(t)| \to -i\infty$ and $|v_i(t)| \to -i\infty$. In this region both $|v_r(t)|$ and $|v_i(t)|$ are imaginary since they correspond to faster than light motion. The Π'_L Reality transformation must be applied to these values to obtain the physical values of the real and imaginary velocities.

The transition from sublight to superluminal particle motion is best understood from the expression for γ:

$$\gamma = [1 + (\alpha^2 - \delta^2)/c^2]^{\frac{1}{2}} \tag{E.18}$$

that follows from eqs. E.9 and E.10. Note that α increases with time t while δ is constant. We now consider the sublight case and then the superluminal case.

Sublight Motion ($|v_r(t)| < c$)
       In this case, supposing the motion begins at a sublight speed, γ begins as a small real quantity (assuming δ is sufficiently small) and increases with time. If $|v_i(t)| = 0$, then as $|v_r(t)| \to c$ we see $\gamma|v_r(t)| = \alpha \to \infty$ and $\gamma \to \infty$. If $|v_i(t)| \neq 0$, then as $|v_r(t)| \to c$ we see $\gamma|v_r(t)| = \alpha$ is finite as is γ. However from eqs. E.9 and E.10 we see that $\gamma \to \infty$ and $\alpha \to \infty$ at

$$|v_r(t)| = [1 + |v_i|^2/c^2]^{\frac{1}{2}} \tag{E.19}$$

Thus the real part of the velocity can exceed the speed of light. The real part of the speed makes γ singular when eq. E.19 is satisfied. The real part of the velocity is limited by the value of the imaginary part of the velocity. If the imaginary part of the velocity is increasing then the limit on the real part is correspondingly increasing.

Superluminal motion
       If

$$|v_r(t)| > [1 + |v_i|^2/c^2]^{\frac{1}{2}}$$

then γ has an imaginary value and consequently the particle's momentum is imaginary as well. As noted earlier the application of Reality group transformation $\Pi'_L = diag(e^{i\chi}, 1, 1, 1)$ (eq. B.38) to the particle 4-momentum with a suitable choice of χ will make the energy E(t) and 3-momenta real physical numbers.
       As time increases, $\alpha \to \infty$, $v \to \infty$, the energy $E \to 0$, and the momentum $|p| \to m$. Thus increasing the velocity well beyond $[1 + |v_i|^2/c^2]^{\frac{1}{2}}$ has a vastly different effect on a particle's energy and

momentum then in the sublight case. The behavior of a particle under a constant electric force is analogous to the case of an accelerating starship that we discussed in appendix D.

E.3 Superluminal Particle Acceleration under a Constant Magnetic Force

We will now consider a superluminal particle traveling through a constant magnetic field. The Lorentz force equation is

$$d\mathbf{p}/dt = qe\mathbf{v} \times \mathbf{B}/c \qquad (E.20)$$

if we decompose the velocity and momentum into components parallel and perpendicular to B:

$$\mathbf{v} = \mathbf{v}_r + i\mathbf{v}_i = \mathbf{v}_\perp + \mathbf{v}_\| = \mathbf{v}_{\perp r} + \mathbf{v}_{\| r} + i\mathbf{v}_{\perp i} + i\mathbf{v}_{\| i} \qquad (E.21)$$

$$\mathbf{p} = \mathbf{p}_r + i\mathbf{p}_i = \mathbf{p}_\perp + \mathbf{p}_\| = \mathbf{p}_{\perp r} + \mathbf{p}_{\| r} + i\mathbf{p}_{\perp i} + i\mathbf{p}_{\| i} \qquad (E.22)$$

then eq. E.20 becomes

$$d\mathbf{p}_\|/dt = 0 \qquad (E.23)$$

$$d\mathbf{p}_\perp/dt = qe(\mathbf{v}_{\perp r} + i\mathbf{v}_{\perp i}) \times \mathbf{B}/c \qquad (E.24)$$

Eq. E.22 implies $\mathbf{p}_\|(t)$ is a constant. Taking real and imaginary parts of eq. E.24 yields

$$d\mathbf{p}_{\perp r}/dt = qe|\mathbf{v}_{\perp r}||\mathbf{B}|\mathbf{u}/c \qquad (E.25)$$

$$d\mathbf{p}_{\perp i}/dt = qe|\mathbf{v}_{\perp i}||\mathbf{B}|\mathbf{w}/c \qquad (E.26)$$

where **u** and **w** are unit vectors in the directions of $\mathbf{v}_{\perp r} \times \mathbf{B}$ and $\mathbf{v}_{\perp i} \times \mathbf{B}$ respectively. Thus in a manner similar to the purely real velocity case we find the superluminal particle is deflected both in its real and imaginary motion in a plane perpendicular to the magnetic field **B**.

Taking the inner product of eq. E.25 with $\mathbf{p}_{\perp r}$ and eq. E.26 with $\mathbf{p}_{\perp i}$ we see that

$$d(\mathbf{p}_{\perp r} \cdot \mathbf{p}_{\perp r})/dt = d(\mathbf{p}_{\perp i} \cdot \mathbf{p}_{\perp i})/dt = 0 \qquad (E.27)$$

Thus the magnitudes of the real and imaginary perpendicular components of the velocity and momentum are constant, and remain equal to their initial values, in a constant magnetic field. Eq. E.27 implies

$$d(\gamma \mathbf{v}_{\perp r})^2/dt = d(\gamma \mathbf{v}_{\perp i})^2/dt = 0 \qquad \text{(E.28)}$$

or

$$\begin{aligned} \gamma |\mathbf{v}_{\perp r}| &= c_r \\ \gamma |\mathbf{v}_{\perp i}| &= c_i \\ |\mathbf{v}_{\perp r}|/|\mathbf{v}_{\perp i}| &= v_r/v_i = c_r/c_i \end{aligned} \qquad \text{(E.29)}$$

where c_r and c_i are constants. As a result the magnitudes $v_{\perp r}$ and $v_{\perp i}$ are constants and equal to their initial values $v_{\perp r0}$ and $v_{\perp i0}$ respectively.

Next we note that eq. E.4 implies

$$\mathbf{v}_{\parallel r} \cdot \mathbf{v}_{\parallel i} + \mathbf{v}_{\perp r} \cdot \mathbf{v}_{\perp i} = 0 \qquad \text{(E.30)}$$

If $\mathbf{v}_{\parallel r} = 0$ (The initial real motion of the particle is perpendicular to the magnetic field **B**.) or $\mathbf{v}_{\parallel i} = 0$ then

$$\mathbf{v}_{\perp r} \cdot \mathbf{v}_{\perp i} = 0 \qquad \text{(E.31)}$$

and the particle executes a circular motion in the plane perpendicular to **B** (assumed to point in the positive z direction.)

We can then parameterize the real and imaginary perpendicular components of $\mathbf{v}_\perp$ by

$$\mathbf{v}_{\perp r} = v_{\perp r}(\hat{x} \cos\theta_r + \hat{y} \sin\theta_r) \qquad \text{(E.32)}$$
$$\mathbf{v}_{\perp i} = v_{\perp i}(-\hat{x} \sin\theta_i + \hat{y} \cos\theta_i) \qquad \text{(E.33)}$$

where by eq. E.31

$$\theta_i = \theta_r \pm \pi \qquad \text{(E.34)}$$

resulting in

$$\mathbf{v}_{\perp i} = v_{\perp i}(\hat{x} \sin\theta_r - \hat{y} \cos\theta_r) \qquad \text{(E.35)}$$

If we extract the $\hat{x}$ components of the real and imaginary parts then the real and imaginary parts of the velocity have the ratio

$$v_{\perp r}\cos\theta_r/(v_{\perp i}\sin\theta_r) = (c_r/c_i)\cot\theta_r \qquad (E.36)$$

Thus we can change the ratio of the real and imaginary parts of a particle's velocity with a constant magnetic field. This could, of course, be useful in piloting a starship. For example, at $\theta_r = 0$ the real part of the perpendicular velocity in the $\hat{y}$ direction has the magnitude 0 and the imaginary part of the perpendicular velocity has the magnitude $v_{\perp i0}$. At $\theta_r = \pi/2$ the real part of the perpendicular velocity in the $\hat{y}$ direction has the magnitude 0 and the imaginary part of the perpendicular velocity has the magnitude $v_{\perp i}$. So if the particle was extracted in the $\hat{y}$ direction (or some other direction) one could adjust the real and imaginary parts of the particle velocity.

Since the magnitudes, $v_{\perp r}$ and $v_{\perp i}$, are constant in time eq. E.25 becomes

$$\gamma d[(\hat{x}\cos\theta_r + \hat{y}\sin\theta_r)]/dt = qe(|\mathbf{B}|/c)[-\hat{y}\cos\theta_r + \hat{x}\sin\theta_r] \qquad (E.37)$$

yielding[67]

$$\theta_r = \theta_{r0} - qe[|\mathbf{B}|/(\gamma c)]t \qquad (E.38)$$

Also since $v_{\perp r}$ and $v_{\perp i}$, are constant in time the particle trajectory in the plane will be a spiral with a linear time dependence:

$$r_r = r_{r0} + v_{\perp r0}t \qquad (E.39)$$

$$r_i = r_{i0} + (c_i/c_r)v_{\perp r0}t \qquad (E.40)$$

where r_r and r_i are the radial distances from the center of the coordinate system.

---

[67] As the particle velocity becomes extremely large θ_r changes more and more rapidly since $\gamma^{-1} \sim v/c$ as $v \to \infty$.

The radii of curvature of the real and imaginary motion can be shown to be

$$\rho_r = cp_{\perp r}/(|q|e|\mathbf{B}|) \qquad\qquad (E.41)$$
$$\rho_i = cp_{\perp i}/(|q|e|\mathbf{B}|) \qquad\qquad (E.42)$$

from eqs. E.25 and E.26.

E.4 Manipulating the Real and Imaginary Parts of a Superluminal Particle's Acceleration and Velocity: Gluon Forces

The discussion in the previous subsections show that an electric field can change the real and imaginary parts of a particle's velocity and a uniform, constant magnetic field causes rotation of the real and imaginary parts of a particle's motion. The combination of these fields can be used for acceleration and the adjustment of the ratio of the real to the imaginary velocities.

Appendix F. Seeing and Navigating through the Cosmos on Superluminal Starships

This appendix is mathematical in nature. Those not mathematically inclined can skip this appendix. The non-mathematical reader need only know that the view of space on faster than light starships is severely distorted but can be remedied by computer imaging techniques.

The view of the universe that a starship crew sees, when the starship is traveling faster than the speed of light, is very different from the view of a spaceship traveling at lower speeds of a few tens of miles per second.

As Mallove (1989) points out[68] an observer on a starship traveling at a relativistic speed near, but below, the speed of light will see the visible stars and galaxies compressed to within a cone in the direction of the starship (Fig. F.1). The cone gets more narrow as the speed of light is approached due to aberration and in the limit as the speed approaches the speed of light becomes a point directly ahead of the starship.

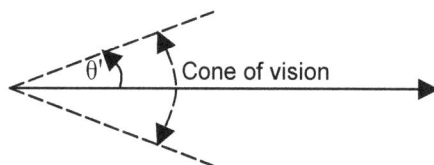

Figure F.1. Cone of visibility around direction of starship motion in the starship coordinate system with the angle θ' determined by eq. F.1 for sublight starship speeds.

---

[68] pp. 182-185. They reference the scientific papers that are the basis of his description.

The relativistic equation for aberration is

$$\cos \theta' = (\cos \theta + \beta)/(1 + \beta \cos \theta) \qquad \text{(F.1)}$$

where θ is the angle of a star (or galaxy) relative to the starship's direction of motion as measured in the earth coordinate system and θ' is the angle of a star (or galaxy) relative to the starship's direction of motion as measured in the starship's coordinate system.
      The inverse relation is

$$\cos \theta = (\cos \theta' - \beta)/(1 - \beta \cos \theta') \qquad \text{(F.2)}$$

F.1 Sublight Case: β < 1
        As $\beta \to 1$ (the speed of light) eq. F.1 indicates $\theta' \to 0°$ showing the entire view of the universe is visually compressed to the exactly forward direction. Fig. F.1 shows the cone of visibility for a spaceship traveling near the speed of light at perhaps .6c - .9c. The cone angle θ' satisfies

$$\cos \theta' > \beta \qquad \text{(F.3)}$$

The rest of the field of view of the starship is total blackness except the point in the directly rearward direction ($\theta' = 180°$) for any object at $\theta = 180°$.

F.2 Superluminal Case: β > 1
        For $\beta > 1$ eqs. F.1 and F.2 still hold and there is a cone of visibility similar to that depicted in Fig. F.1. However the cone angle θ' for superluminal speeds, $\beta > 1$, satisfies the relation

$$\cos \theta' > 1/\beta \qquad \text{(F.4)}$$

The rest of the field of view of the starship is total blackness, as in the sub-light speed case, except the point in the directly rearward direction ($\theta' = 180°$). We note that as β gets very large the cone of visibility becomes larger. At $\beta = \infty$ the cone of visibility becomes the angular region between $\theta' = 0°$ and $\theta' = 90°$ (the forward hemisphere).

F.3 Superluminal Starship Visibility

As a result visual navigation at high superluminal speeds becomes difficult although one can conceive of electronic imaging that "undoes" the effects of aberration and enables visual navigation.

A further problem is the location of a destination. If we send a starship from the earth to a star, for example 30 light years away, we have to project the location of the star at the time the starship arrives based on the star's current motion as determined by earth observation. If the motion of the star is modified by the gravitation effects of other nearby stars during the 30 years that the light from the star was traveling to earth, or if the star's motion is not accurately determined, a starship could arrive at a point that is some distance from the star.

Thus navigation to a destination is a significant issue.

F.4 Effect of Doppler Shift at Superluminal Speeds

A starship traveling at relativistic sublight speeds will see stars having their color changed significantly due to the Doppler Shift effect. At superluminal speeds the Doppler Shift will also change the colors of objects "seen" by starship occupants.

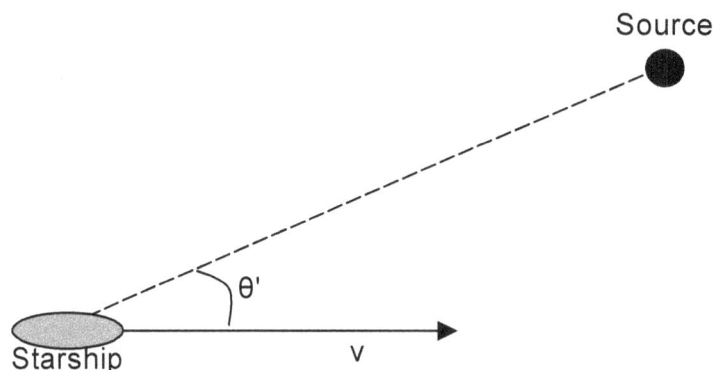

Figure F.2. The angle of a source θ' with respect to the starship's velocity **v**.

This issue is again surmountable if we use electronic imaging techniques to "undo" the Doppler shift and thus display stars as they normally look in the visible

The relativistic Doppler shift for sublight speeds of a light wave of frequency v is given by

$$v = v_0(1 - \beta^2)^{\frac{1}{2}}/(1 - \beta \cos \theta') \qquad (F.5)$$

where v_0 is the frequency of the light emitted by the source and θ' is the angle of the source relative to the starship's velocity (Fig. F.2). The Doppler shift for superluminal speeds is

$$v = v_0(\beta^2 - 1)^{\frac{1}{2}}/(\beta \cos \theta' - 1) \qquad (F.6)$$

This can be seen by considering an electromagnetic plane wave, which is a combination of

$$\cos[(k \cdot x - vt)/2\pi] \quad \text{and} \quad \sin[(k \cdot x - vt)/2\pi] \qquad (F.7)$$

Upon transforming from the earth coordinate system, for example, to a coordinate system moving in the x-direction at a speed faster than light (See eqs.) both the energy (v' up to a constant) and the time t' obtain a factor of i (that cancel each other) so eq. F.6 is the correct frequency in the superluminal frame. The sign of the frequency is always positive by convention due to the form of electromagnetic waves and eq. F.4 dictates the form of the denominator in eq. F.6.

For large $\beta \gg 1$ eq. F.6 becomes approximately

$$v \approx v_0/\cos \theta' \qquad (F.8)$$

In the forward direction $\theta' = 0$ the Doppler shift goes to zero. Due to eq. F.4 the maximum value of the Doppler shift for large β in the field of vision is

$$v \approx \beta v_0 \qquad (F.9)$$

So the "wide" angle electromagnetic waves are shifted to large frequency.

Eq. F.6 and the discussion that follows suggests that frequency shifts will be substantial for extremely fast starships. The result will be a distorted view of the universe.

However electronic imaging techniques can be implemented to restore the view that humans would normally see. Thus the combined effects of aberration and the Doppler shift on the view of space from the starship bridge can be electronically corrected to give a "normal" view of space.

Appendix G. Quark-Gluon Fluid Experiments and Accelerator Thrust

This appendix is mathematical in nature. Those not mathematically inclined can skip this appendix. The non-mathematical reader need only know that quark-gluon plasmas are being created in laboratory experiments using colliding lead, gold and uranium ions, that it is possible to collide spherules of matter (small spheres of matter weighing less than milligram), and that the collision fragments can be controlled by electromagnets.

G.1 Relativistic Heavy Ion Collisions

The collision of highly relativistic heavy ions such as lead-lead, gold-gold, and uranium-uranium collisions is a relatively new field in experimental particle/nuclear physics. Experiments at the SPS (CERN) collide gold and lead ions at an energy of 17 GeV per nucleon, and experiments at RHIC (Brookhaven National Laboratory) collided lead with lead and gold with gold ions at 130 GeV per nucleon. At the time of this writing initial ion collision experiments have begun at LHC (CERN) using colliding lead ions[69] – each with an energy of 1380 GeV per nucleon. All of these experiments indicate something new and exciting is happening in highly relativistic, heavy ion collisions that do not happen in p-p (proton-proton) collisions. A colliding ion pair temporarily creates a fireball containing quarks and gluons in a macroscopic state known as a *perfect fluid*. Within this region the quark-gluon fluid acts as if it had little or no viscosity (friction).[70] After a short time,[71] 5 – 7 fm/c, during which this perfect fluid is explosively expanding, the fluid enters

---

[69] About 400 nucleons collide with each other in lead-lead and gold-gold collisions.

[70] The particles in the quark-gluon fluid rapidly reach thermal equilibrium after an ion collision through an as yet undetermined mechanism.

[71] Time is measured in fm/c = 3.34×10^{-24} sec. The symbol fm represents a distance of one fermi = 10^{-15} m. The constant c is the speed of light.

the "Freeze-Out" state dispersing into numerous particles that stream out of the collision region.

During the time that the quark-gluon perfect fluid exists it is in thermal equilibrium and the laws of thermodynamics apply as well as the equations of ideal relativistic hydrodynamics. Many papers[72] now exist in the literature, which describe the nature and evolution of the macroscopic quark-gluon fluid as well as the freeze out state that follows. These papers as well as the experimental results that can be expected from LHC studies in the next few years will provide a sound basis for the understanding of the dynamics of the quark-gluon state and the subsequent freeze out state.

The details of these results are not of immediate interest for our goal of designing a starship engine. Rather we want to investigate an extension of the present experimental LHC configuration that can serve as a prototype for a starship drive. In brief we view the drive as composed of an intersecting storage ring(s) setup that collides ionized heavy elements (in small spheres – spherules) in a chamber defined by confining laser or particle beams, or by a magnetic bottle, that is similar in concept to a rocket combustion chamber and nozzle. The macroscopic fireball generated by the collision(s) contains a large number of "unconfined" quarks with complex velocities that are funneled by the beams (and/or magnets) into complex rocket thrust for the starship. The major design details of a starship engine appear in appendix H.

A large amount of data has been collected at the SPS and RHIC. For our purposes the data of interest can be summarized as:

1. The generated fireball is a macroscopic, hot, dense, strongly interacting perfect fluid containing unconfined quarks and gluons with complex spatial momentum.

---

[72] See E. Shuryak, arXiv:0807.3033 (2007) and references therein.

2. Data[73] from RHIC include:
    Maximum diameter of fireball for lead at various
    energies is about 5 fm.
    Initial size of fireball – a diameter of 0.8 fm for lead.
    Transverse v/c at edge of fireball is about .5 for times τ:
    3 – 9 fm/c.

3. Some Estimates:[74]

- Hydrodynamic relaxation time are below 1 fm/c.
- Expansion rate – "Hubble fireball constant" τ^{-1} = $\partial_\mu u^\mu$ where u^μ is the fluid 4-velocity. τ is estimated to be of the order of 1 to several fm/c.
- Fireball lifetime – Hubble fireball time until freeze out 5 – 7 fm/c.
- Equilibration time (time for hydrodynamic fireball stage to begin) 0.6 fm/c.
- Initial fireball volume expansion is linear in t but grows to t^3.
- Fireball energy density 25 GeV/fm^3 at fireball center at 130 GeV.
- Ellipticity of a fireball typically is a 10% - 20% effect.
- The transition to quark-gluon perfect fluid occurs about 6.5 times cold nuclear density at 0.14 nucleons/fm^3.
- The energy density in the fireball center rises roughly linearly with GeV/nucleon: density (GeV/ fm^3) = 0.19×(GeV/nucleon).
- The time to cross a nucleus of radius R ≅ $1.2A^{1/3}$ is approximately 6 fm/c for lead or uranium.

---

[73] K. Dusling and D. Teaney, arXiv; 0710.5932 (2007).
[74] P. F. Kolb and U. Heinz, arXiv:nucl-th/0305084 (2003)

4. LHC – ALICE Group Results[75]

The ALICE group lead-lead experiment[76] at LHC found that the elliptic flow of charged particles at the LHC energy of 2760 GeV/nucleon was 30% greater than that encountered at the RHIC where the energy was 130 GeV/nucleon. The quark-gluon fluid at LHC energies also displays little friction and thus is a perfect fluid. Future studies at the LHC are expected to yield extremely precise data on the nature and behavior of the quark-gluon fluids.

G.2 Spherule Accelerators

In the previous section we considered hadron-hadron collisions for the case of heavy ions such as lead, gold and uranium that consist only of a nucleus since each ion is stripped of all its electrons. The radius of a heavy ion nucleus such as lead or uranium is about 6 fm. We now wish to consider spherule-spherule accelerators and collisions since we wish to maximize the mass expelled per second of the starship thrust.

The first issue is the magnet strength needed to confine spherules to a "circular" orbit realizing that there are many additional magnetic ensembles needed to stabilize the orbits of cycling spherules in a cyclotron.[77] Another issue that we will not address is the mechanism to accelerate and inject spherules into the main synchrotron ring.

G.2.1 Hadron Collider "Turning" Magnets

The magnitude |**B**| of the magnetic field required to maintain a radius of curvature ρ is given by

$$|\mathbf{B}| = pc/(|q|\rho) \qquad\qquad (G.1)$$

---

[75] CERN Courier **51**, 7 (April, 2011).

[76] K. Aamodt et al, Phys. Rev. Lett. **105**, 252302 (2011).

[77] See Lee (2004) and other accelerator books.

where p is the particle's momentum and q is the particle's charge. The ratio of the required magnetic field for a spherule $|\mathbf{B_s}|$ to the magnetic field $|\mathbf{B_h}|$ required for a hadron ion to have the same radius of curvature ρ is

$$|\mathbf{B_s}|/|\mathbf{B_h}| = (p_s/p_h)(|q_h|/|q_s|) \tag{G.2}$$

Assuming the hadron and ion velocities are equal, then eq. G.2 becomes

$$|\mathbf{B_s}|/|\mathbf{B_h}| = (m_s/m_h)(|q_h|/|q_s|) \tag{G.3}$$

where m_s and m_h are the masses of the spherule and ion respectively. If the spherule has a mass of 0.01 mg and is composed of U-238 then its mass ratio with a single U-238 ion is the number n of uranium atoms in the spherule. The charge of a fully stripped U-238 ion is 92e. If one electron is stripped from each U-238 atom (on average) in a .01 mg spherule then the charge on the spherule is 0.34 μC (micro-Coulombs). The ratio of the required spherule accelerator magnetic field to the ion's magnetic field in this case is 92 times the ion's magnetic turning field

$$|\mathbf{B_s}|/|\mathbf{B_h}| = n(92/n) = 92 \tag{G.4}$$

or, more generally, for an element of atomic number A

$$|\mathbf{B_s}|/|\mathbf{B_h}| = A \qquad ` \tag{G.4a}$$

Thus the turning magnetic field strength for a spherule is the element's atomic number times that required for an ion of the same element. Magnets are possible with a field strength of about 100 times that of the LHC turning magnets.

G.2.2 Electric Acceleration of Spherules

RF generators accelerate particles[78] in a synchrotron using intense electric fields. The process may be simply viewed as an acceleration of a particle with a change in energy E given by the change in potential energy in the rf generator in the amount

$$q\Delta V$$

where ΔV is the change in electric potential energy. The ratio of the energy gain per atom/ion of the spherule to an individual ion is

$$(q_s/q_h)/n = (n/92)/n = 1/92 \qquad (G.5)$$

or, more generally, for an element of atomic number A

$$(q_s/q_h)/n = 1/A \qquad (G.5a)$$

since we assume one electron is removed per atom in the spherule and each ion has 92 electrons stripped from it. Thus the energy gain per atom in the spherule is 1/92 that of an ion under the same potential difference.

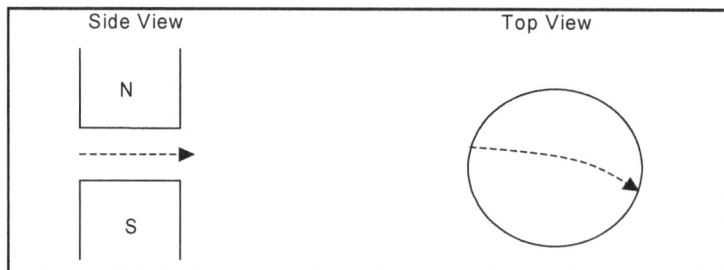

Figure G.1. Path (dashed line) of a charged particle between the cylindrical N and S poles of a magnet: a side view showing the particle is not deflected vertically, and a top view showing the particle is deflected horizontally to the right.

---

[78] Both the real and imaginary parts of particle velocities.

G.2.3 Spherule Synchrotron Radiation

Accelerating charged particles radiate energy. In the case of a synchrotron of radius ρ the power radiated is

$$P = 2q^2e^2c\ \beta^4 / [3\rho^2(1-\beta^2)^2] \qquad (G.6)$$

When β ~1 (relativistic)

$$P \approx [2q^2e^2c/(3\rho^2)]\ (E/mc^2)^4 \qquad (G.7)$$

Since the mass ratio of a u-238 ion to a 0.01 mg spherule is 0.5×10^{-15}, the power radiated per atom by an accelerating spherule at a given energy is much less than that of a single U-238 ion:

$$P_s/P_h \approx (q_s/q_h)^2(m_h/m_s)^4 = 1/(92n)^2 \qquad (G.8)$$

or for an element of atomic number A and an n atom spherule

$$P_s/P_h \approx 1/(An)^2 \qquad (G.8a)$$

In the case of a 0.01 mg U-238 spherule the power radiated by the spherule is a factor of 2.7×10^{-35} less than the power radiated by a U-238 ion. Thus the power loss is very much smaller for spherules.

G.2.4 Spherule Synchrotrons are Possible Today

Although turning magnets must be much more powerful, and electric rf generators must also be much more powerful, spherule accelerators are viable with present technology or modest extensions thereof. The much lower rate of synchrotron radiation is a very favorable factor.

Appendix H. Superluminal Starship Drives

This appendix is mathematical in nature. Those not mathematically inclined can skip this appendix. The non-mathematical reader need only know that faster than light starship drives are possible. It also describes drives that use the imaginary quark-gluon plasma thrust to cause a starship to rotate.

H.1 Types of Starships

There are two general types of starships due to the nature of starship drives: starships based on circular accelerators and starships based on linear accelerators. In both cases we assume the process that creates the quark-gluon plasma thrust is a stream of collisions of spherules of some material.

We therefore begin with a consideration of the spherule collision process. We then proceed to describe the features of some general forms of starships.

H.2 Colliding Spherule Starship Thrust

Our starship drive design is based on the R&D efforts expended to develop fusion power reactors using Inertial Confinement Fusion and on the extensive experience gained in building ultra-high energy colliding beam accelerators.[79] The steps of the engine cycle are:

---

[79] Since the mathematical analysis of the starship propulsion mechanism requires complex computer codes that have yet to be developed we shall describe starship propulsion in qualitative terms except for a few major points concerning rotating starships with artificial gravity. A detailed mathematical investigation would require a significant programming effort, and extensive experimental input, that is beyond the current state of the art.

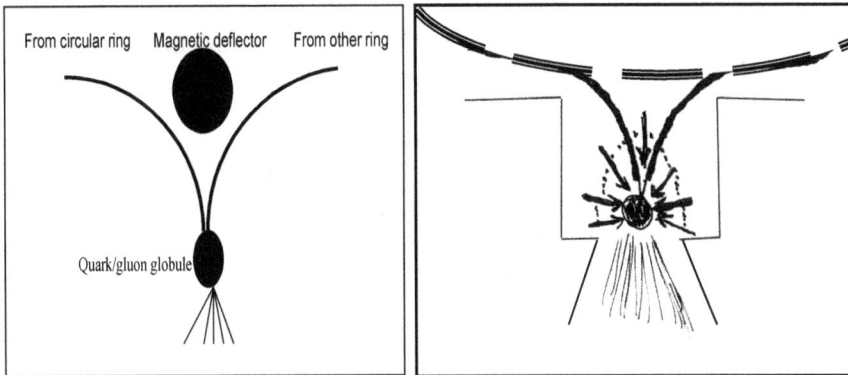

Figure H.1a. Diagram of the collision region of the intersecting spherules. The spherules (thick lines) that are extracted from the accelerator rings are thicker until they are near the point of intersection where each spherule is compressed by high power lasers or particle beams (not shown) to nuclear density. They then collide, generating a fireball that is constrained from expanding by multiple beams, except towards the rear from which the fireball of quark-gluon plasma streams to generate the starship thrust. The straight lines enclosing the "combustion chamber" indicate the enclosing sides of the starship hull. The arrows represent some of the laser or particle beams (or magnetic field pressure) used to enclose the fireball allowing expansion only towards the rear. The "dots" between the arrows indicate that there is likely to be an array of many beams confining the expanding fireball to expansion to the rear of the starship as shown.

1. Highly charged spherules are accelerated in "standard" colliding rings with new, more powerful magnets and rf accelerator modules. Two streams of spherules are diverted by magnets into the collision module as illustrated by Fig. H.1. They are bent from their circular trajectory *to almost the thrust direction but retain their momentum component towards collision*. One thinks here of a collision at about a 45° angle.

   A starship using linear accelerators to accelerate spherules would orient the linear accelerators to meet at the collision point. An angle of about 45° between the linear accelerators would appear to be the best for maximizing the thrust produced. The remaining items in this section apply to both circular and linear accelerators.

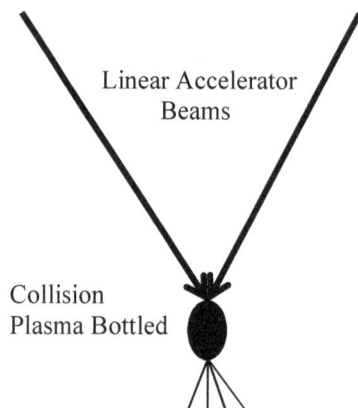

Linear Accelerator
Beams

Collision
Plasma Bottled

Figure H.1b. Colliding spherule linear accelerator beams produce a quark-gluon globule that provides the thrust.

2.   About-to-collide spherules are compressed to nuclear density in the shape of thin ellipsoids by sets of laser beams similar to those used for inertial confinement fusion. The long axis of each spheroid is oriented in the "thrust" direction to give a maximal collision surface for the colliding spherules. Each of the colliding spherules are compacted to ultra-high density as if being prepared for fusion (at perhaps twice nuclear density) using an array of powerful lasers or particle beams focused on each spherule. After compaction each pair of two spherules collide in a few fm/c to create a macroscopic fireball. The lasers/particle beams may vaporize part or all of a spherule but the ultrahigh density of the resulting compact stream makes that issue irrelevant.

3.   Colliding   fireballs   in   the   multi-femtometer   size   starship "combustion" chamber are confined to the chamber by laser or particle beams or by powerful magnets on the sides, and also in front, of such strength as to only allow the fireballs to exit to the rear   providing   a   quark-gluon   thrust.   The   strength   of   the confinement beams, or magnets, must be orders of magnitude

stronger than beams currently used in Inertial Confinement Fusion (ICF) since the density and explosive force is so much more than in ICF devices. The complex velocities of the quarks within the quark-gluon thrust lead to complex starship velocities that can exceed the speed of light.[80]

4. The thrust should have a high complex velocity. The thrust will accelerate the starship to a speed well in excess of the speed of light. Since the exhaust speed is so high, small amounts of matter in the thrust will cause the starship speed to increase rapidly to high velocities.

5. The energy of the colliding spherule rings should be as high as possible to produce maximal thrust velocity.

The success of the starship engine requires carefully synchronized events at the fm/c level. The effort expended in developing this starship engine will also result in advances in ICF reactors for fusion power and in elementary particle accelerators for deeper studies of elementary particle physics and cosmology. Thus the R&D payoff will be very substantial.

H.3 Fireball Complex Spatial Momentum "Bubble" in Real Space-Time

A fireball created by the high energy collision of heavy ions or heavy atom spherules is substantially different from the collision region in nucleon-nucleon collisions. Fig. H.2 shows a visualization of the fireball. Inside the fireball, quarks and gluons form a perfect fluid and have complex spatial momenta. Outside the fireball the normal real-valued spatial dimensions prevail.

---

[80] A fireball expanding to the rear due to the confining effect of beams, or magnets, assumes an ellipsoid shape as it exits from the starship. The real and imaginary parts of the quark velocities can be directed to the "rear." Consequently, the rearward component of the quark velocities would be complex-valued.

A fireball is created by the collision of heavy ions,[81] and rapidly (in a fm/c or so) becomes a perfect fluid described by thermodynamics. The fireball explosively expands as described in section H.2. After 6 – 7 fm/c or so, its energy density becomes low enough to enter the "freeze out" stage where the quarks and gluons in the residue of the fireball combine (including quark-antiquark pairs created from the vacuum) to produce "normal particles" such as pions, protons , neutrons, and so on.

In the case of colliding spherules a much larger fireball will be produced if the colliding spherules are first compacted to nuclear density. Each compacted spherule can then be viewed as a "super-nucleus."

The surface of a fireball has a certain surface tension – otherwise it would not be approximately spherical/ellipsoidal in shape. It does have some ellipticity at the 15% - 20% level. However the speed of fireball expansion shows the surface tension is much less than the pressure of expansion and thus may be ignored to leading order.

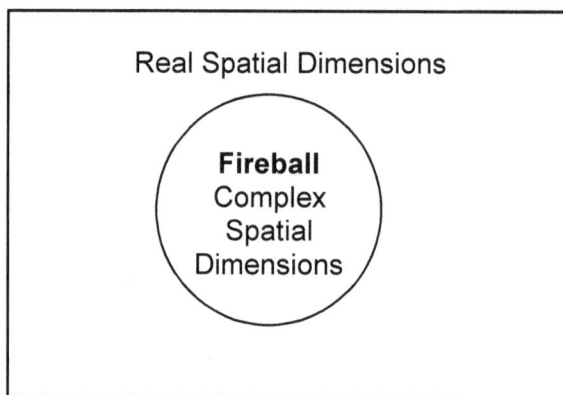

Real Spatial Dimensions

Fireball
Complex
Spatial
Dimensions

Figure H.2. A depiction of a quark-gluon fireball in which the quarks and gluons have complex spatial momenta. The fireball is a complex "bubble" within the real space of our experience.

---

[81] Other remnants of the colliding ions are also produced.

The existence of surface tension, and the freeze out of ion-ion collision fireballs after 6 – 7 fm/c, do raise the issue of whether the starship thrust would experience a drag due to these effects.

H.4 Freeze Out Stage – Impact on Starship Thrust

The laser or particle beams defining the "combustion chamber" sides confine the expanding fireball to expand only to the rear of the starship and thus provide thrust. An expanding fireball's perfect fluid elongates as shown in Fig. H.3 and the fluid explosively expands out of the rear of the chamber. In a short time period of perhaps a few fm/c freeze out occurs and the fireball "dissipates" with the quarks and gluons combining (together with quark-antiquark pairs excited from the vacuum) to transform into "normal" elementary particles.

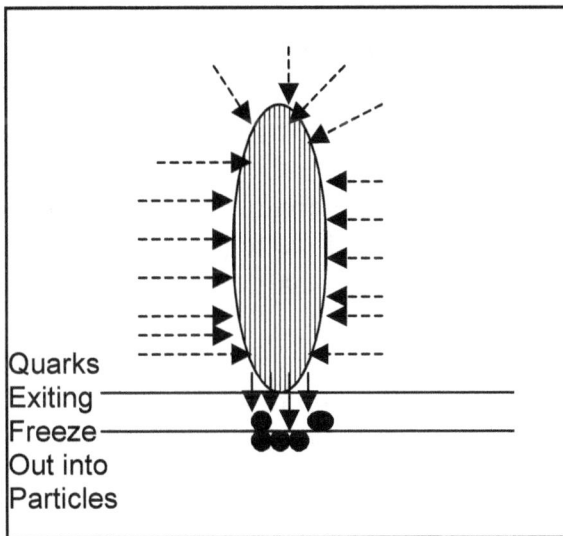

Figure H.3. Diagram of the expanding, beam guided fireball exiting the "combustion" chamber. Dotted line arrows represent the confining laser or particle beams (or magnetic field). The emerging quark-gluon fluid (indicated by the first horizontal line) dissipates (the freeze out) and forms "normal" elementary particles (indicated by the second horizontal line).

One may ask if the freeze out, which represents the breakup of the fireball surface, causes a "drag" to occur that reduces the thrust.[82] It appears that the quark thrust, which has a complex velocity, is the true thrust of the engine. The later recombination of the quarks and gluons outside the engine produces normal particles with real velocities. The imaginary part of the quark velocities is subsumed within the produced normal particles. This process of thrust transforming to another form occurs in chemical rockets as well. The molecules in the chemical rocket thrust, which often contains some free radicals, often recombine in the tail of the rocket thrust in a manner analogous to freeze out. The chemical recombination in the thrust tail does not affect rocket performance.

H.5 Starship Engine Energy Source

The starship engine that we have designed requires massive amounts of energy. At this point in time the only feasible energy source for starships is nuclear energy. It is reasonable to expect that fusion energy, a more concentrated energy source, will become a reality within the next thirty years. In either case the starship will need an energy source to drive the spherule accelerator rings and associated devices of the engine for periods up to perhaps a few months or years, then turn off for perhaps many years, and then resume operations for further maneuvers.

In the extreme case of travel to another galaxy, the energy source will need to turn off for up to millions of years of starship time. While the energy source is turned off, a residual "battery" will need to operate to support monitoring the progress of time, activating the startup of the main energy source, and possibly to detect and monitor objects ahead of the starship in the line of flight. This battery source may well be a plutonium source similar to those used in current space probes.[83]

---

[82] Similar drag effects occur in fluid dynamics near surfaces.

[83] We note that a natural nuclear reactor existed in the Congo Region of Central Africa for millions of years. (Parenthetical note: Could this be the stimulus for the rapid evolution of species in Africa including very early Mankind?)

The main energy source, if it is a nuclear reactor of some kind, will probably have to be a reactor that is different from current nuclear reactors. Chapter 5 describes long shelf life nuclear reactors that could meet this need. Since the startup process from a battery driven state needs to be gradual due to a "small" battery, it appears the set of nuclear reactors would be composed of perhaps five reactors of increasing size. The battery starts the smallest reactor by concentrating its nuclear fuel. The smallest reactor then generates the energy to concentrate that fuel, and start the second smallest reactor, and so on until the main reactor(s) is started. At this point the accelerators power up and starship thrust begins.

If the source of the energy is fusion energy then the startup process might begin in small stages in the boot up of the fusion reaction through fusing larger and larger amounts of (perhaps) ^{3}He with increasingly powerful laser beams a la the tokamak approach.

During the coasting period of a starship, nuclear reactors should be powered down to conserve nuclear fuel. When powering down a nuclear reactor power source the nuclear material (U^{235} or plutonium) (the reactor fuel) residing in the medium of the reactors would be diluted to sharply reduce fission reactions to "near zero" using the energy of the next largest reactor. The smallest reactor would be powered down by a battery. This battery would retain enough energy to bring the smallest reactor back up after the coasting period ends. Then the reactors would boot up in turn to provide energy to the vehicle. (The battery would be at extremely low temperature during a coasting phase and thus not lose a significant amount of electrical power.)

In the case of a fusion power source a battery could be used to initiate the fusion power. A gradual turnoff process could execute at the start of a coasting phase to bring the fusion process to zero in such a way that the battery could initiate the boot up process for the fusion power source at the end of a coasting period.

H.6 An Alternate Starship Accelerator Ring Engine Design

In an earlier work Blaha (2009b) we proposed an alternate accelerator ring design for quark-gluon fluid acceleration. It appears that this design is not feasible with current or near term (one hundred years) technology. The reasons are:

1. The quark-gluon fluid ring is not feasible because fireballs cannot be injected and accelerated to form a ring in a few fm/c - the time available.

2. A quark-gluon fluid ring would be similar to fusion tokamaks but much more challenging in its requirements for confinement and stability due to the much higher density and temperature of a quark-gluon fluid ring.

H.7 Varieties of Complex Thrust

There are a variety of possible configurations for faster than light starships. Chapter 9 gave a qualitative view of some possible starship designs. In this appendix we categorize the starship configurations by their thrusters.

H.7.1 Rear Thrust Exhausts on Circular Accelerator Starships

In Blaha (2010a), and later books, quarks and gluons were shown to have complex 3-momenta. The real part of the 3-momenta of each quark was orthogonal to the imaginary part of its 3-momenta.

If we "add" pairs of exiting quarks using the mechanism of Fig. H.1a to create a complex 3-momentum to the rear as in Fig. H.4, then a complex total pair 3-momentum is created that generates a rearward thrust.

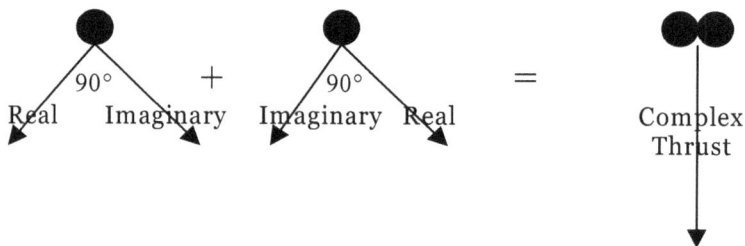

Figure H.4. The sum of two quarks' momenta to give a complex total 3-momenta (thrust) directed to the rear.

The complex thrust can be used in a one thrust exhausts, or two thrust (or multi-thrust) exhausts as shown in Figs. H.5 and H.6.

Figure H.5. A circular accelerator starship with one thrust exhaust.

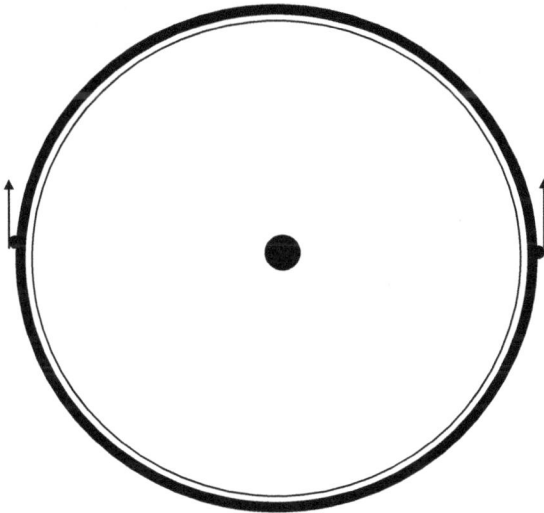

Figure H.6. Top view of a circular starship with two complex thrust exhausts.

These types of starships generate thrust that only moves the starship forward in space. They do not cause the starship to rotate. The next subsection introduces the use of the imaginary perpendicular part of a starship's thrust to cause the starship to rotate.

H.7.2 Rotating Circular and Cylindrical Starships due to Imaginary Part of Thrust

In the case of disc-shaped and cylindrical starships the imaginary part of the quark-gluon thrust can be directed by magnets to be tangent to the circular edge of the starship. The resulting tangential force causes the starship to rotate through an imaginary angle. The rotation creates a *negative* centripetal force in the starship – "artificial gravity" that varies with the distance from the central axis of the starship. Fig. H.7 shows a "top view" of a starship.

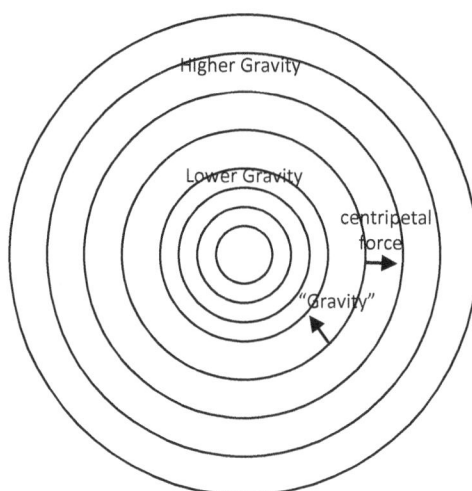

Figure H.7. Top View of a Disc-like or Cylindrical Starship. View of cargo/people levels of inner hub. The center has lower "gravity." Circles indicate levels of equal artificial gravity in a disc-shaped or cylindrical starship. The outer parts have higher "gravity." The "gravity" force is inward towards the center on all levels. The centripetal force is outward from the center as shown by the arrow in the diagram and the discussion of eq. H.5 below (opposite to the direction of conventional centripetal force.) The artificial "gravity" force is thus inward to the center.

Fig. H.8 is an example of a disc-shaped starship with the thrust direction displayed. The horizontal thrust arrows represent the real part of the quark-gluon thrust. It accelerates the starship to the left in the figure. The vertical arrows represent the imaginary part of the thrust. It rotates the starship in a counterclockwise direction.

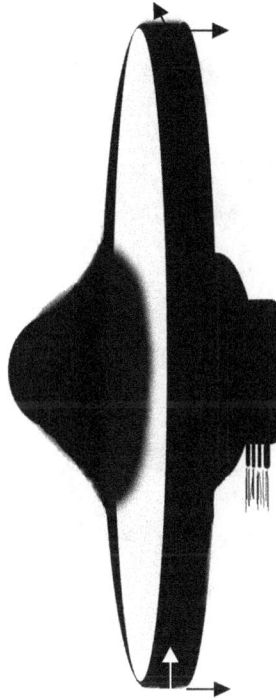

Figure H.8. A disc-like starship with thrust consisting of a real and imaginary part. The real part drives the ship to the left. The imaginary part is tangent to the edge of the starship causing it to rotate counterclockwise.

H.7.2.2 Artificial Gravity Levels - Rotating Circular and Cylindrical Starships

In rotating starships the real part of the thrust propels a starship. The combination of the real and imaginary parts of the thrust enables a starship to exceed the speed of light. The imaginary part of the thrust causes the starship to spin around its central axis.

   We will examine the case of a cylindrical starship and see how the "artificial gravity" emerges as a result of the imaginary part of the thrust. Fig. H.9 shows a cylindrical starship and Fig. H.10 shows the cylindrical coordinate system used to calculate its motion.

Figure H.9. Cigar shaped starship with "horizontal accelerator ring(s). The real part of the thrust points downward. The imaginary part of the thrust is horizontal and tangent to the cigar surface. The fins are for supplementary nuclear maneuvering rockets.

The force generated by the quark-gluon thrust has the form

$$\mathbf{F} = g_r \check{\mathbf{z}} + i g_i \mathbf{\emptyset} \qquad (H.1)$$

in the starship's cylindrical coordinates rest frame where ž is a unit vector in the positive z direction and ∅ is a unit vector in the positive ∅ angle direction. g_r and g_i are constants specifying the thrust. g_r is the starship's mass times its real-valued acceleration in the ž direction. g_i is the starship's mass times its imaginary-valued acceleration in the ∅ direction. The time derivative of the momentum **p** of a small part of mass m_0 at the edge of the starship is

$$dp/dt = m_0(\gamma z')'\check{z} + m_0[(\gamma\rho\varnothing')' + \gamma\rho'\varnothing')]\varnothing + m_0[-\gamma\rho(\varnothing')^2 + (\gamma\rho')']\rho \qquad (H.2)$$

where **ρ** is the unit vector in the positive ρ direction and the ´ symbol indicates a time derivative.

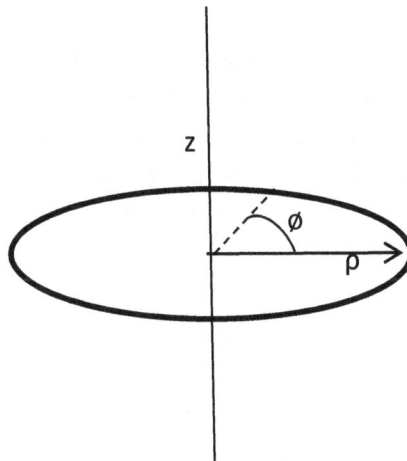

Figure H.10. Starship cylindrical coordinate system. The cigar shaped Starship in Fig. H.9 is centered on the z axis.

If we set ρ´= 0 since there is no force component in the **ρ** direction and the starship is not contracting, then setting

$$F = dp/dt \qquad (H.3)$$

implies

$$m_0\rho(\gamma\phi')' = ig_i \tag{H.4}$$

The centripetal force is

$$F_{centripetal} = m_0\gamma v_c^2/\rho \tag{H.5}$$

where v_c is the rotational velocity of the mass. It satisfies

$$v_c = \rho\phi' \tag{H.6}$$

Since ϕ' is imaginary by eq. H.4, v_c Is imaginary and thus the centripetal force $F_{centripetal}$ is negative by eq. H.5. The centripetal force is "opposite" to the centripetal force normally experienced – away from the center. The artificial "gravity" force – really the opposite of the centripetal force – is towards the center. See Fig. H.7.

The artificial "gravity" engendered by the rotation raises an issue. The acceleration of the rotation implied by eqs. H.2 and H.3, if continued, would generate an enormous inward "gravitational" force crushing the starship's occupants. This problem can be solved by causing an oscillation in the rotation between clockwise and counter-clockwise directions by repeatedly flipping the value of the imaginary force g_i (eq. H.1) to maintain a constant rotation speed (and thus have constant values of "gravity.")

A circular (disc-shaped) starship can also rotate to create artificial gravity. Fig. 9.5 illustrates a rotating disc starship, Again the gravitational force is towards the center with lower gravity in the central region. The disc moves to the left due to the real thrust.

H.7.3 Starships Based on Linear Particle Accelerators

Starships based on linear accelerators can have a variety of forms. Fig. 9.6 contains a diamond shaped starship. Generally starships based on linear accelerators need great length – of the order of miles for the primary linear accelerator tubes unless a very rapid linear accelerator mechanism is developed. The angles between the linear accelerator tubes are determined by maximizing the efficiency, and amount, of thrust generation. Thus the shape of the *lower half* of the diamond-shaped starship is more or less determined. The lower part of the starship can be sharp edged diamond shaped or rounded diamond shaped. The shape of the upper part is not determined unless there is a substantial problem with space dust. In that case the upper part of the starship should minimize the effects of space dust.

H.8 Benefits of starship R&D – Fusion Power and Advanced Particle Accelerators

The Research and Development of a starship will be a long (30 – 100 years), expensive effort that will be best done by a consortium of technologically advanced nations including the USA, United Kingdom, China, Japan, India and The European Union. An important motivation for this effort is its impact on the research and development efforts for fusion power as well as particle accelerator development.

Figure H.11 shows the benefits fallout from a starship development effort as envisioned in this book.

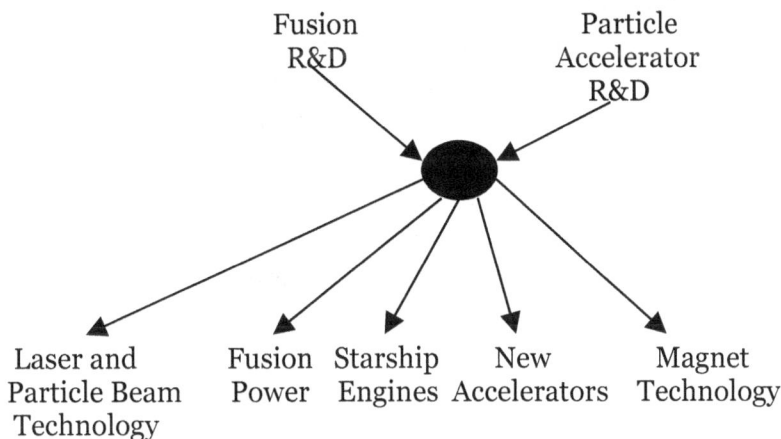

Fusion
R&D

Particle
Accelerator
R&D

Laser and   Fusion Starship   New   Magnet
Particle Beam   Power Engines Accelerators Technology
Technology

Figure H.11. Benefits from a starship R&D effort accrue in other areas such as fusion power – a vital requirement for the future and new particle accelerators to advance human knowledge.

Appendix I. Faster Than Light Kinetic Theory and Thermodynamics

This appendix is mathematical in nature. Those not mathematically inclined can skip this appendix. The non-mathematical reader need only know that quark-gluon globules used for starship thrust and composed of faster than light particles in thermodynamic equilibrium have thermodynamic and kinetic theory behavior generally similar to globules of slower than light particles in thermodynamic equilibrium.

This appendix contains extracts from chapter 40 of Blaha (2012a) plus some additional material. It shows that faster than light physics is in many ways similar to slower than light physics in the important areas of kinetic theory and thermodynamics. It is particularly important in the study of faster than light quark-gluon plasmas. These plasmas are known to come rapidly to thermodynamic equilibrium after they are created in spherule-spherule or ion-ion collisions.

The reason for the rapid transition to thermodynamic equilibrium is not experimentally known at present. We believe it is due to extremely rapid collisions of the quark-parton constituents of nucleons at faster than light speeds. After thermodynamic equilibrium is achieved then the kinetic theory and thermodynamics discussed in this appendix, and originally in Blaha (2012a), is applicable. We will describe our superluminal kinetic theory and the resulting Thermodynamics. There are strong similarities with non-relativistic Thermodynamics but also significant differences.

I.1 Superluminal Kinetic Theory

Assemblages of large numbers of particles embody the Maxwell-Boltzmann distribution. The Boltzmann H theorem is the beginning point for derivations of the non-relativistic Maxwell-Boltzmann distribution. The non-relativistic Maxwell-Boltzmann distribution has the form

$$f(\mathbf{v}, \mathbf{r}) = n(m/(2\pi kT))^{3/2} \exp\{-[m(\mathbf{v} - \mathbf{v}_0)^2/2 + V(r)]/(kT)\} \qquad (I.1)$$

where n is the particle density, T is the temperature, $\mathbf{v}_0$ is the average velocity, m is the particle mass, V(r) is an external conservative force, and k is Boltzmann's constant. We can express the Maxwell-Boltzmann distribution as In terms of a Hamiltonian

$$H(\mathbf{v}, \mathbf{r}) = m\mathbf{v}^2/2 + V(r) \qquad (I.2)$$

as

$$f(\mathbf{v}, \mathbf{r}) = n(m/(2\pi kT))^{3/2} \exp\{-H(\mathbf{v} - \mathbf{v}_0, \mathbf{r})/(kT)\} \qquad (I.3)$$

I.1.1 Relativistic Form of the Maxwell-Boltzmann Distribution

If we assume that we have a container containing a distribution of relativistic (sublight) particles with an average velocity $\mathbf{v}_0 = 0$, and no external force, then the form of eq. I.3 generalizes to the relativistic Maxwell-Boltzmann distribution

$$f_R(\mathbf{v}) = C_R \exp\{-H/(kT)\} \qquad (I.4)$$

where C_R is a normalization constant and H is the relativistic hamiltonian for a free particle:

$$H = c(m^2c^2 + \mathbf{p}^2)^{1/2} \qquad (I.5)$$

with $\mathbf{p} = \gamma m\mathbf{v}$ and

$$\gamma = (1 - v^2/c^2)^{-1/2} \qquad (I.6)$$

I.1.2 Superluminal Form of the Maxwell-Boltzmann Distribution

The superluminal form of Maxwell-Boltzmann distribution is based on the form of the mass shell condition for superluminal particles:

$$E^2 - c^2 \mathbf{p}^2 = m^2 c^4 \tag{I.7}$$

which implies a free hamiltonian

$$H_S = c(\mathbf{p}^2 - m^2 c^2)^{\frac{1}{2}} \tag{I.8}$$

where

$$\mathbf{p} = \gamma_s m \mathbf{v} \tag{I.9}$$

and

$$\gamma_s = (v^2/c^2 - 1)^{-\frac{1}{2}}$$

The seemingly slight differences between eqs. I.8, I.9, and eq. I.5 causes some major differences between superluminal and relativistic kinetic theory and thermodynamics. Relativistic kinetic theory and thermodynamics are qualitatively similar in many ways with their non-relativistic counterparts.

One major difference is the behavior of kinematic variables near the speed of light:

As $v \to c$ Below the Speed of Light

$$p \to \infty$$
$$H \to \infty$$

As $v \to c$ From Above the Speed of Light

$$p \to \infty$$
$$H_S \to \infty$$

As $v \to \infty$

$$p \to mc$$
$$H_S \to 0$$

Thus as v ranges from c to ∞, H_S decreases monotonically from ∞ to zero and p decreases from ∞ to mc. This behavior contrasts with H in eq. I.5, which increases monotonically with p as v increases from 0 to c. Thus the sublight Maxwell-Boltzmann distribution decreases with v as v increases from 0 to c.

The superluminal Maxwell-Boltzmann distribution *increases* with v as v increases from c to ∞ as we see below. The superluminal Maxwell-Boltzmann distribution decreases with p as p increases from mc to ∞. *As a result the natural physical parameterization of the Maxwell-Boltzmann distribution should be in terms of the momentum rather than the velocity.* Thus Boltzmann's H function which normally is

$$H_B(t) = \int d^3v \, f(\mathbf{v}, t) \log f(\mathbf{v}, t)$$

must be replaced with[84]

$$H_{BS}(t) = \int d^3p \, f_S(\mathbf{p}, t) \log f_S(\mathbf{p}, t)$$

The equilibrium superluminal Maxwell-Boltzmann distribution can be derived from $H_S(t)$. It has the same general form as the relativistic distribution

$$f_S(\mathbf{p}) = C_S \exp\{-H_S/(kT)\} \qquad (I.10)$$

where C_S is a normalization constant and H_S is the superluminal hamiltonian for a free particle.

We now apply the normalization condition[85]

$$n = N/V = \int d^3p f_S(\mathbf{p}) = C_S \int d^3p \, \exp\{-H_S/(kT)\} \qquad (I.11)$$

---

[84] Note the additional factor of m^3 in $\int d^3p$ will be absorbed in the normalization (eq. I.11).

[85] We note that using $\int d^3v$ rather than $\int d^3p$ in eq. I.11 would result in a divergence – another reason for our choice of integration parameter.

where n is the particle density, N is the number of particles in the system, and V is the volume of the system. We calculate C_S by evaluating the integral:

$$n = 4\pi C_S \int_{mc}^{\infty} dp \ p^2 \ \exp\{-H_S/(kT)\} \qquad (I.12)$$

Letting $x = p/(mc)$ and $\alpha = mc^2/(kT)$ we see eq. I.12 becomes

$$n = 4\pi m^3 c^3 C_S \int_{1}^{\infty} dx \ x^2 \ \exp\{-\alpha(x^2 - 1)^{\frac{1}{2}}\} \qquad (I.13)$$

Then letting $y^2 = x^2 - 1$ we find

$$n = 4\pi m^3 c^3 C_S \int_{0}^{\infty} dy \ y(y^2 + 1)^{\frac{1}{2}} \ \exp(-\alpha y)$$

$$= -m^3 c^3 C_S G^{31}_{13}(\alpha^2/4 \mid {}^{\ 0}_{-3/2, 0, \frac{1}{2}}) \qquad (I.14)$$

where $G^{31}_{13}(...)$ is Meijer's G-Function.[86] Therefore

$$C_S = -[m^3 c^3 G^{31}_{13}((mc^2/(2kT))^2 \mid {}^{\ 0}_{-3/2, 0, \frac{1}{2}})/n]^{-1} \qquad (I.15)$$

The most probable momentum of a particle p_p is the maximum of

$$p_p = \text{Max}\{p^2 \exp[-H_S/(kT)]\}$$

$$= \{(2(kT)^2/c^2)[1 + (1 - m^2c^4/(kT)^2)^{\frac{1}{2}}]\}^{\frac{1}{2}} \qquad (I.16)$$

For large T or small T the maximum is

$$p_p \approx 2kT/c \ > mc$$

---

[86] See Gradshteyn (1965) integral 3.389.2 and p. 1068 for the properties of Meijer's G-Function.

The velocity v_p corresponding to the maximum in the momentum is

$$v_p = cp_p/(p_p^2 - m^2c^2)^{1/2}$$

For large T or small T, the velocity v_p corresponding to the maximum in the momentum is approximately

$$v_p \approx c + \tfrac{1}{2} m^2c^5/(2kT)^2$$

I.2 Superluminal Thermodynamics

Turning now to the thermodynamics of a dilute superluminal gas implied by the superluminal Maxwell-Boltzmann distribution we begin by calculating the average energy per particle

$$\varepsilon = C_S \int d^3p\, H_S \exp[-H_S/(2kT)] \Big/ \int d^3p\, C_S \exp[-H_S/(kT)] \tag{I.17}$$

$$= (C_S/n) \int d^3p\, H_S \exp[-H_S/(kT)]$$

$$= (C_S/n)2kT\alpha\, 4\pi m^2c^3 \int_0^\infty dy\, y^2(y^2+1)^{1/2} \exp(-\alpha y)$$

$$= -(C_S/n)m^3c^5 G^{31}{}_{13}(\alpha^2/4 \mid^{-1/2}{}_{-2,0,\,1/2})$$

$$= mc^2\, G^{31}{}_{13}((mc^2/(2kT))^2 \mid^{-1/2}{}_{-2,0,\,1/2}) \Big/ G^{31}{}_{13}((mc^2/(2kT))^2 \mid^{0}{}_{-3/2,0,\,1/2}) \tag{I.18}$$

The Maxwell-Boltzmann normalization factor is related to the energy per particle by

$$C_S = -n\varepsilon/(m^3c^5 G^{31}{}_{13}(\alpha^2/4 \mid^{-1/2}{}_{-2,0,\,1/2})) \tag{I.19}$$

Note that C_S is proportional to the energy in contrast to the non-relativistic case where the Maxwell-Boltzmann normalization factor $C = (3m/(4\pi\varepsilon))^{3/2}$.

We now calculate the superluminal pressure for the case of a distribution of superluminal particles bouncing on a wall perpendicular to the z-axis. The wall is assumed to be a perfectly reflecting plane. The pressure is the average force per unit area due to the gas of superluminal particles. The number of particles bombarding the wall per second with $v_z > 0$ is $v_z f_S(\mathbf{p})d^3 p$. Thus the pressure is

$$P = \int d^3 p\, 2 p_z v_z f_S(\mathbf{p}) \tag{I.20}$$

where the particle momentum changes by $2p_z$ due to reflection. Due to spherical symmetry one expects the average values for the various components of $\mathbf{v}$ to be equal. Consequently we can re-express eq. I.20 as

$$P = 1/3 \int d^3 p\, 2 m \gamma_S v^2 f_S(\mathbf{p}) \tag{I.21}$$
$$= 1/3 \int d^3 p\, 2 p v f_S(\mathbf{p})$$

Since

$$v = cp/(p^2 - m^2 c^2)^{\frac{1}{2}} \tag{I.22}$$

we see

$$P = 8\pi c/3 \int_{mc}^{\infty} dp\, p^4 f_S(\mathbf{p})/(p^2 - m^2 c^2)^{\frac{1}{2}}$$

Following steps similar to eqs. I.12 – I.15 lead to

$$P = m^4 c^4 C_S G^{31}_{13}((mc^2/(2kT))^2 |\,^{\frac{1}{2}}_{-2,0,\,\frac{1}{2}}) \tag{I.23}$$

The *equation of state* relating the pressure and energy is

$$P = -(m/c)\{G^{31}_{13}((mc^2/(4kT))^2 |\,^{\frac{1}{2}}_{-2,0,\,\frac{1}{2}})\big/ G^{31}_{13}(\alpha^2/4 \,|^{-\frac{1}{2}}_{-2,0,\,\frac{1}{2}})\}n\varepsilon \tag{I.24}$$

Substituting for ε we find

$$P = -(nm^2 c)\{G^{31}_{13}(\rho \,|\,^{\frac{1}{2}}_{-2,0,\,\frac{1}{2}})\big/ G^{31}_{13}(\rho \,|^{0}_{-3/2,0,\,\frac{1}{2}})\} \tag{I.25}$$

where

$$\rho = (mc^2/(2kT))^2 \tag{I.26}$$

Turning now to the consideration of a dilute gas, the internal energy of the gas can be defined by[87]

$$U(t) = N\varepsilon \qquad (I.27)$$

We note that the work done by the superluminal gas if its volume increases by dV is PdV. Then the superluminal (and usual) form of the first law of thermodynamics is

$$dQ = dU + PdV \qquad (I.28)$$

where Q is the heat absorbed. The heat capacity of the system for constant volume is

$$C_V = (\partial U/\partial T)_V \qquad (I.29)$$

The second law of thermodynamics, Boltzmann's H theorem, is based on

$$H = -S/kV \qquad (I.30)$$

where H is the negative of the entropy divided by k times the volume V. In systems where there are no superluminal particles, the H theorem states that the entropy never decreases for an isolated gas of fixed volume.

We can calculate H for a superluminal system under equilibrium conditions, H_e, from[88]

$$H_e = \int d^3p f_S(\mathbf{p}) \ln(f_S(\mathbf{p})) \qquad (I.31)$$

$$= \int d^3p f_S(\mathbf{p}) [\ln C_S - H_S/(kT)] \qquad (I.32)$$

---

[87] The internal energy of a gas of non-interacting non-relativistic particles is U(t) = 3NkT/2. In the superluminal case it appears that it is eq. I.18.

[88] We consistently assume that integrals over the momentum $\int d^3p$ are the proper integration (rather than integrations over velocity $\int d^3v$) because, for example, the calculation of the normalization constant eq. I.11 would diverge if the integration were over $\int d^3v$.

$$= n \ln C_S - \int d^3 p f_S(\mathbf{p}) H_S/(kT)$$

$$= n \ln C_S - n\varepsilon/(kT) \tag{I.33}$$

by eqs. I.11 and I.17. Therefore

$$S = -kVH_{Se} = -kN \ln C_S + N\varepsilon/T \tag{I.34}$$

Consequently we obtain the superluminal *and* standard non-relativistic result

$$1/T = (\partial S/\partial U)_x \tag{I.35}$$

where x represents all other extensive variables.

I.3   Approximate   Calculation   of   Kinetic   and Thermodynamic Quantities

We can obtain more tractable expressions for kinetic and thermodynamic quantities by assuming $\mathbf{p}^2 \gg m^2 c^2$ and approximating the hamiltonian (eq. I.8) with

$$H_{Sa} = cp \tag{I.36}$$

The approximate normalization condition is

$$n = N/V = \int d^3 p f_{Sa}(\mathbf{p}) = C_{Sa} \int d^3 p \exp\{-H_{Sa}/(kT)\} \tag{I.37}$$

where n is the particle density, N is the number of particles in the system, and V is the volume of the system. C_S is determined by

$$n = 4\pi C_{Sa} \int_{mc}^{\infty} dp \, p^2 \exp\{-cp/(kT)\} \tag{I.38}$$

Letting $\alpha = c/(kT)$ we see eq. I.38 becomes

$$n = 4\pi C_{Sa}\, d^2/d\alpha^2 \int_{mc}^{\infty} dp\, \exp(-\alpha p)$$

$$= 4\pi C_{Sa}\, d^2/d\alpha^2\, [(1/\alpha)\exp(-\alpha mc)] \tag{I.39}$$

Therefore the normalization factor is

$$C_{Sa} = n\big/\{4\pi\, d^2/d\alpha^2\, [(1/\alpha)\exp(-\alpha mc)]\} \tag{I.40}$$

The most probable momentum of a particle p_p is the maximum of

$$p_{pa} = \text{Max}\{p^2\exp[-H_{Sa}/(kT)]\}$$
$$= 2kT/c \tag{I.41}$$

The velocity v_{pa} corresponding to the maximum in the momentum is

$$v_{pa} = cp_{pa}/(p_{pa}^2 - m^2c^2)^{\frac{1}{2}}$$

For large T or small T, the velocity v_{pa} corresponding to the maximum in the momentum is approximately

$$v_{pa} \approx c + \tfrac{1}{2}\, m^2c^5/(2kT)^2$$

Turning now to the thermodynamics implied by the superluminal Maxwell-Boltzmann distribution we begin by calculating the average energy per particle

$$\varepsilon_a = \int d^3p\, H_{Sa}\, \exp[-H_{Sa}/(kT)]\Big/\int d^3p\, \exp[-H_{Sa}/(kT)] \tag{I.42}$$

$$= (C_{Sa}/n) \int d^3p\, H_{Sa}\, \exp[-H_{Sa}/(kT)]$$

$$= -(4\pi c C_{Sa}/n)\, d^3/d\alpha^3 [(1/\alpha)\exp(-\alpha mc)] \to 3kT \quad \text{for } T \gg mc$$

where $\alpha = c/(kT)$.[89]

The Maxwell-Boltzmann normalization factor is related to the energy per particle by

$$C_{Sa} = -n\varepsilon_a/\{4\pi c \ d^3/d\alpha^3[(1/\alpha)\exp(-\alpha mc)]\} \qquad (I.43)$$

Note that C_{Sa} is proportional to the energy ε_a in contrast to the non-relativistic case where the Maxwell-Boltzmann normalization factor $C = (3m/(4\pi\varepsilon))^{3/2}$.

We now calculate the superluminal pressure for the case of a distribution of superluminal particles bouncing on a wall perpendicular to the z-axis. The wall is assumed to be a perfectly reflecting plane. The pressure is the average force per unit area due to the gas of superluminal particles. The number of particles bombarding the wall per second with $v_z > 0$ is $v_z f_{Sa}(\mathbf{p})d^3p$. Thus the pressure is

$$P_a = \int d^3p \, 2p_z v_z f_{Sa}(\mathbf{p}) \qquad (I.44)$$

where the particle momentum changes by $2p_z$ due to reflection. Due to spherical symmetry one expects the average values for the various components of $\mathbf{v}$ to be equal. Consequently we can re-express eq. I.44 as

$$P_a = 1/3 \int d^3p \, 2m\gamma_s v^2 f_{Sa}(\mathbf{p}) \qquad (I.45)$$
$$= 1/3 \int d^3p \, 2pv f_{Sa}(\mathbf{p})$$

Since

$$v = cp/(p^2 - m^2c^2)^{\frac{1}{2}} \qquad (I.46)$$

we see

$$P_a = 8\pi c/3 \int_{mc}^{\infty} dp \, p^4 f_{Sa}(\mathbf{p})/(p^2 - m^2c^2)^{\frac{1}{2}} \qquad (I.47)$$

$$\cong 8\pi c/3 \int_{mc}^{\infty} dp \, p^3 f_{Sa}(\mathbf{p})$$

---

[89] The Superluminal case differs from the non-relativistic case: $\varepsilon_a = 3kT/2$. An example of $\varepsilon_a = 3kT$ is a crystal with a potential energy of compression. See p. 192 Morse (1964).

Evaluating eq. I.47 yields

$$P_a = -(8\pi c/3)\, C_{Sa}\, d^3/d\alpha^3 [(1/\alpha)\exp(-\alpha mc)] \qquad (I.48)$$

The *equation of state* relating the pressure and energy is[90]

$$P_a = 2/3\, n\varepsilon_a \qquad (I.49)$$

For $T \gg mc$ we found[91]

$$\varepsilon_a \rightarrow 3kT \qquad (I.50)$$

Then, contrary to non-relativistic kinetic theory, we find $(T \gg mc)$

$$P_a = 2nkT \qquad (I.51)$$

Turning now to the consideration of a dilute gas the internal energy of the gas for $T \gg mc$ is

$$U(t) = N\varepsilon \rightarrow 3NkT \qquad (I.52)$$

We note again that the work done by the superluminal gas if its volume increases by dV is PdV. Then the superluminal (and usual) form of the first law of thermodynamics is

$$dQ = dU + PdV \qquad (I.53)$$

where Q is the heat absorbed. The heat capacity of the system for constant volume is $(T \gg mc)$

$$C_V \rightarrow 3Nk \qquad (I.54)$$

---

[90] The same equation of state as non-relativistic kinetic theory. See p. 72 Huang (1965).

[91] Later we will define temperature in terms of the entropy S as $1/T = (\partial S/\partial U)_x$ where x is all other extensive variables.

The second law of thermodynamics, Boltzmann's H theorem, is based on

$$H = -S/kV \qquad (I.55)$$

where H is the negative of the entropy divided k times the volume V. In systems where there are no superluminal particles, the H theorem states that the entropy never decreases for an isolated gas of fixed volume.

We can calculate H_{BS} for a superluminal system under equilibrium conditions, H_{BSea}, from[92]

$$H_{BSea} = \int d^3 p f_{Sa}(\mathbf{p}) \ln(f_{Sa}(\mathbf{p})) \qquad (I.56)$$

$$= \int d^3 p f_{Sa}(\mathbf{p})[\ln C_{Sa} - H_{Sa}/(kT)] \qquad (I.57)$$

$$= n \ln C_{Sa} - \int d^3 p f_{Sa}(\mathbf{p}) H_{Sa}/(kT)$$
$$= n \ln C_{Sa} - n\varepsilon_a/(kT) \qquad (I.58)$$

by eqs. I.11 and I.17. Therefore

$$S_a = -kVH_{BSea} = -kN \ln C_{Sa} + N\varepsilon_a/T \qquad (I.59)$$

The superluminal *and* standard non-relativistic result still holds

$$1/T = (\partial S/\partial U)_x \qquad (I.60)$$

where x represents all other extensive variables.

---

[92] We have consistently assumed that integrals over the momentum $\int d^3 p$ are the proper integration (rather than integrations over velocity $\int d^3 v$) because, for example, the calculation of the normalization constant eq. D.11 would diverge if the integration were over $\int d^3 v$.

I.4 Superluminal Kinetics and Thermodynamics are Similar to the Non-Relativistic Case

In the previous sections we have shown that kinetic theory and the laws of thermodynamics are different but somewhat similar in the superluminal, relativistic, and non-relativistic cases. Recently due to a stream of new data on quark-gluon plasmas from various laboratories around the world, theorists have been analyzing the thermodynamics and dynamics of this data.

In view of the very real likelihood that quarks are faster than light particles, theoretical analyses and data fits should be done using the superluminal theory presented here as well as traditional approaches.

REFERENCES

Blaha, S., 2004, *Quantum Big Bang Cosmology: Complex Space-time General Relativity, Quantum Coordinates, Dodecahedral Universe, Inflation, and New Spin 0, ½, 1 & 2 Tachyons & Imagyons* (Pingree-Hill Publishing, Auburn, NH, 2004).

_____, 2006, *A Unified Quantitative Theory Of Civilizations and Societies: 9600 BC - 2100 AD* (Pingree-Hill Publishing, Auburn, NH, 2006)

_____, 2007a, *Physics Beyond the Light Barrier: The Source of Parity Violation, Tachyons, and A Derivation of Standard Model Features* (Pingree-Hill Publishing, Auburn, NH, 2007).

_____, 2007b, *The Origin of the Standard Model: The Genesis of Four Quark and Lepton Species, Parity Violation, the ElectroWeak Sector, Color SU(3), Three Visible Generations of Fermions, and One Generation of Dark Matter with Dark Energy* (Pingree-Hill Publishing, Auburn, NH, 2007).

_____, 2008, *A Complete Derivation of the Form of the Standard Model With a New Method to Generate Particle Masses SECOND EDITION* (Pingree-Hill Publishing, Auburn, NH, 2008)

_____, 2009a, *Bright Stars, Bright Universe* (Pingree-Hill Publishing, Auburn, NH, 2009)

_____, 2009b, *To Far Stars and Galaxies: Second Edition of Bright Stars, Bright Universe* (Pingree-Hill Publishing, Auburn, NH, 2009).

_____, 2010a, *The Standard Model's Form Derived from Operator Logic, Superluminal Transformations and GL(16)* (Pingree-Hill Publishing, Auburn, NH, 2010).

_____, 2010b, *SuperCivilizations: Civilizations as Superorganisms* (McMann-Fisher Publishing, Auburn, NH, 2010).

_____, 2011c, *All the Universe! Faster Than Light Tachyon Quark Starships & Particle Accelerators with the LHC as a Prototype Starship Drive Scientific Edition* (Pingree-Hill Publishing, Auburn, NH, 2011).

_____, 2011d, *From Asynchronous Logic to The Standard Model to Superflight to the Stars* (Blaha Research, Auburn, NH, 2011).

_____, 2012a, *From Asynchronous Logic to The Standard Model to Superflight to the Stars volume 2: Superluminal CP and CPT, U(4) Complex General Relativity and The Standard Model, Complex Vierbein General Relativity, Kinetic Theory, Thermodynamics* (Blaha Research, Auburn, NH, 2012).

_____, 2012b, *Standard Model Symmetries, and Four and Sixteen Dimension Complex Relativity; The Origin of Higgs Mass Terms* (Blaha Research, Auburn, NH, 2012).

Bussard, R., DeLauer, R. 1958, *Nuclear Rocket Propulsion* (McGraw-Hill, New York, 1958).

Bussard, R., 1965, *Fundamentals of Nuclear Flight* (McGraw-Hill, New York, 1965).

Dewar, James, 2008, *To The End Of The Solar System: The Story Of The Nuclear Rocket* (2nd ed.), (Apogee Books, 2008).

Freeman, Marsha, 2009, *Krafft Ehricke's Extraterrestrial Imperative* (Apogee Books, 2009).

Huang, K., 1965, *Statistical Mechanics* (Joh Wiley & Sons, New York, 1965).

Huang, K., 1992, *Quarks, Leptons & Gauge Fields Second Edition* (World Scientific, River Edge, NJ, 1992).

Huang, K., 1998, *Quantum Field Theory* (John Wiley, New York, 1998).

Lee, S. Y., 2004, *Accelerator Physics Second Edition* (World Scientific Publishing co, New Jersey, 2004).

Mallove, E. F. and Matloff, G. L., (1989) *The Starflight Handbook* (John Wiley, New York, 1989).

Morse, P. M., (1964) *Thermal Physics* (W. A. Benjamin, New York, 1964).

Schmidt, S. and Zubrin, R. (eds.), 1996, *Islands in the Sky* (John Wiley, New York, 1996).

Weinberg, S., 1972, *Gravitation and Cosmology* (Wiley, New York, 1972).

Zubrin, R., 2000, *Entering Space* (Penguin Putnam, New York, 2000).

INDEX

About the Author

Stephen Blaha is an internationally known physicist with interests in Science, the Arts, and Technology. He had an Alfred P. Sloan Foundation scholarship in college. He received his Ph.D. in Physics from Rockefeller University. He has served on the faculties of several major universities. He was also a Member of the Technical Staff at Bell Laboratories, a manager at the Boston Globe Newspaper, a Director at Wang Laboratories, and President of Blaha Software Inc and of Janus Associates Inc. (NH).

Among other achievements he was a co-discoverer of the "r potential" for heavy quark binding developing the first (and still the only demonstrable) non-abelian gauge theory with an "r" potential; first suggested the existence of topological structures in superfluid He-3; first proposed Yang-Mills theories would appear in condensed matter phenomena with non-scalar order parameters; first developed a grammar-based formalism for quantum computers and applied it to elementary particle theories; first developed a new form of quantum field theory without divergences (thus solving a major 60 year old problem that enabled a unified theory of the Standard Model and Quantum Gravity without divergences to be developed); first developed a formulation of complex General Relativity based on analytic continuation from real space-time; first developed a generalized non-homogeneous Robertson-Walker metric that enabled a quantum theory of the Big Bang to be developed without singularities at t = 0; first generalized Cauchy's theorem and Gauss' theorem to complex, curved multi-dimensional spaces; received Honorable Mention in the Gravity Research Foundation Essay Competition in 1978; first developed a physically acceptable theory of faster-than-light particles; first showed a universe with three complex spatial dimensions is icosahedral; first derived a composition of extrema method in the Calculus of Variations; first quantitatively suggested that inflationary periods in the history of the universe were not needed; first proved Gödel's Theorem implies Nature must be quantum; provided a new alternative to the Higgs Mechanism, and Higgs particles, to generate masses; first showed how to resolve logical paradoxes including Gödel's Undecidability Theorem by developing Operator Logic and Quantum Operator Logic; first developed a quantitative harmonic oscillator-like model of the life cycle, and interactions, of civilizations; first showed how equations describing superorganisms also apply to civilizations; and first developed an axiomatic derivation of the forms of The Standard Model with WIMPs from geometry – space-time properties – The faster than light Standard Model.

He has had a major impact on a succession of elementary particle theories: his Ph.D. thesis (1970), and papers, showed that quantum field theory calculations to all orders in ladder approximations could not give scaling deep inelastic electron-nucleon scattering. He later showed the eigenvalue equation for the fine structure constant α in Johnson-Baker-Willey QED had a zero at $\alpha = 1$ not 1/137 by solving the Schwinger-Dyson equations to all orders in an approximation that agreed with exact results to 8th order in α thus ending interest in this theory. In 1979 at Prof. Ken Johnson's (MIT) suggestion he calculated the proton-neutron mass difference in the

MIT bag model and found the result had the wrong sign reducing interest in the bag model. These results all appear in Physical Review papers. In the 2000's he repeatedly pointed out the shortcomings of SuperString theory and showed that The Standard Model's form could be derived from space-time geometry by an extension of Lorentz transformations to faster than light transformations. This deeper space-time basis greatly increases the possibility that it is part of THE fundamental theory.

In the early 1980's Blaha was also a pioneer in the development of UNIX for financial, scientific and Internet applications: benchmarked UNIX versions showing that block size was critical for UNIX performance, developing financial modeling software, starting database benchmarking comparison studies, developing Internet-like UNIX networking (1982) and developing a hybrid shell programming technique (1982) that was a precursor to the PERL programming language. He was also the manager of the AT&T ten-year future products development database. His work helped lead to commercial UNIX on computers such as Sun Micros, IBM AIX minis, and Apple computers.

In the 1980's he pioneered the development of PC Desktop Publishing on laser printers. and was nominated for three "Awards for Technical Excellence" in 1987 by PC Magazine for PC software products that he designed and developed.

In the past ten years Dr. Blaha has written over 35 books on a wide range of topics. Some recent major works are: *From Asynchronous Logic to The Standard Model to Superflight to the Stars, All the Universe!* and *SuperCivilizations: Civilizations as Superorganisms.*

www.ingramcontent.com/pod-product-compliance
Lightning Source LLC
Chambersburg PA
CBHW080236270326
41926CB00020B/4262